78th Annual Eastern Snow Conference 2022

Online
1 – 8 June 2022

ISBN: 978-1-7138-6653-4

Printed from e-media with permission by:

Curran Associates, Inc.
57 Morehouse Lane
Red Hook, NY 12571

Some format issues inherent in the e-media version may also appear in this print version.

Printed with permission by Curran Associates, Inc. (2023)

For permission requests, please contact Eastern Snow Conference
at the address below.

Eastern Snow Conference
C/O Dr. Krystopher Chutko
117 Science Pl-Dept. Geography
Saskatoon, Sk, Canada S7N 5C8

https://www.easternsnow.org/

Additional copies of this publication are available from:

Curran Associates, Inc.
57 Morehouse Lane
Red Hook, NY 12571 USA
Phone: 845-758-0400
Fax: 845-758-2633
Email: curran@proceedings.com
Web: www.proceedings.com

Proceedings of the
78th ANNUAL

EASTERN SNOW CONFERENCE

Joint Congress with Canadian Meteorological and Oceanographic Society and Canadian Geophysical Union
1-8 June 2022

Virtual Meeting

ISSN: 0424-1932

Proceedings of the 78th Annual Eastern Snow Conference

CONTENTS

Session #1: ESC General Session

Session #3: Monitoring and Modeling Cryospheric Change

Session #5: Observation and Modeling of Snow Processes: Integrating Legacy and New Tools to Advance Snow Science

FOREWORD

This proceedings volume contains papers presented at the 78th Eastern Snow Conference (ESC) held 1-8 June 2022 as a virtual joint meeting between the ESC, the Canadian Meteorological and Oceanographic Society (CMOS), and the Canadian Geophysical Union (CGU). The meeting featured sessions on the observation and modeling of snow processes, field measurement of solid precipitation, new approaches to snowpack measurement and general monitoring of the cryosphere, snow and ice in mountainous and Arctic regions, and snowpack properties.

The ESC is a joint United States and Canadian forum for discussing recent work on operational, applied and scientific issues related to snow and ice. It also retains an increasing interest as a symposium where novel approaches to cryospheric science of international significance are presented. The ESC has published an annual series of proceedings since 1952. Typical topics include studies of snow and ice as materials, snow removal, meteorological forecasting, river ice control, snow hydrology, snow chemistry, glaciology, remote sensing of snow and ice, and snow ecology. Membership in the ESC is open to all interested individuals and corporations. Additional copies of the current proceedings and all back issues can be obtained from the Secretary. Additional information about the Eastern Snow Conference may be found at: http://www.easternsnow.org/.

We thank members of the ESC executive committee for their continued work toward making each annual meeting a success. This includes, but is not limited to, the work of the Research Committee who evaluate all student presentations and recommend awardees for the three student prizes.

The virtual meeting was organized jointly by CMOS, CGU, and ESC. Individual sessions were supported by a team of volunteers who managed the online services and recordings. Special thanks to Andrew Ireson and Francis Zwiers for leading the Local Arrangements and Scientific Program committees, respectively.

The 2022 meeting of the Eastern Snow Conference and these Proceedings were made possible in part by Corporate membership from Campbell Scientific Canada, Geonor, Gradient Wind Engineering, and Hoskin Scientific Ltd. Printing of the Proceedings was supported by the Centre for Hydrology, University of Saskatchewan.

Campbell Scientific (Canada) Corp.
Edmonton, Alberta, Canada
http://www.campbellsci.ca

Hoskin Scientific Limited
Burnaby, British Columbia, Canada
http://www.hoskin.ca

GEONOR
Augusta, New Jersey, USA
http://www.geonor.com

Gradient Wind Engineering
Ottawa, Ontario, Canada
www.gradientwind.com

Centre for Hydrology
University of Saskatchewan
Saskatoon/Canmore, SK/AB
Canada
https://research-groups.usask.ca/hydrology/

We look forward to seeing you at the 79th Eastern Snow Conference meeting to be held 6-8 June 2023 at the Nurture Nature Center in Easton, Pennsylvania, USA.

Krystopher Chutko and Craig Smith
ESC Proceedings Co-Editors
University of Saskatchewan and Environment and Climate Change Canada

STATEMENT OF PURPOSE

The Eastern Snow Conference (ESC) is a joint Canadian/US organization founded in the 1940s, originally with members primarily from eastern North America. Our current members are scientists, snow surveyors, engineers, technicians, professors, students, and operational and maintenance professionals from North America, the United Kingdom, Japan, and Germany. There is a western counterpart to the ESC, the Western Snow Conference (WSC), which also is a joint Canadian/US organization.

The Eastern Snow Conference is a forum that brings the research and operations communities together to discuss recent work on scientific, applied, and operational issues related to snow and ice. The location of the conference alternates yearly between the United States and Canada, and attendees present their work by either giving a talk or presenting a poster. Most resulting papers are reviewed, edited, and published in our yearly *Proceedings of the Eastern Snow Conference*. In recent years, the ESC meetings have included sessions on snow physics, winter survival of animals, snow and ice loads on structures, river ice, remote sensing of snow and ice, and glacier processes. Volumes of the *Proceedings* can be found in libraries throughout North America and Europe, and the papers are also available through the National Technical Information Service (NTIS) in the United States and CISTI in Canada. All volumes of the Proceedings are now available at *easternsnow.org*.

*   *   *   *   *   *   *   *   *

Le Colloque sur la neige région Est (ESC) est une organisation américaine-canadienne fondée dans les années '40 et dont les membres provenaient à l'origine principalement de l'est de l'Amérique-du-Nord. Actuellement, les membres, qu'ils soient chercheurs, techniciens, ingénieurs, professeurs, étudiants, et spécialistes des services d'exploitation et d'entretien, viennent non seulement d'Amérique-du-Nord, mais aussi du Royaume-Uni, du Japon, et d'Allemagne. Le Colloque sur la neige-région Ouest (WSC), aussi une organisation américaine-canadienne, est l'homologue de l'ESC pour l'ouest nord-américain.

L'ESC est un forum qui rassemble chercheurs et responsables des services d'exploitation pour discuter des travaux récents sur les problèmes scientifiques, opérationnels, ou autres dus à la neige et à la glace. Le site de cette réunion annuelle alterne entre les États Unis et le Canada. Les participants y présentent les résultats de leurs travaux par des communications orales ou au moyen d'affiches. Ces communications, une fois revues et éditées, sont publiées dans les *Annales de l'ESC*. Dans les années récentes, les réunions de l'ESC ont inclus des sessions sur la physique de la neige, la survie hivernale de la faune, les forces exercées par la neige et la glace sur les structures et les bâtiments, la glace de rivière, la télédétection de la neige et de la glace, et les processus glaciaires. Les *Annales de l'ESC* sont accessibles dans la plupart des bibliothèques scientifiques d'Amérique-du-Nord et d'Europe. Des copies d'articles peuvent être obtenues du National Technical Information Service (NTIS) aux États Unis et son équivalent au Canada, le CISTI. Toutes les volumes des *Annales de l'ESC* sont maintenant disponibles sur *easternsnow.org*.

EXECUTIVES FOR THE 2021-22 78th EASTERN SNOW CONFERENCE

PRESIDENT
Krystopher Chutko
Saskatoon
Saskatchewan, Canada

PAST PRESIDENT
Eli Deeb
Hanover
New Hampshire, USA

VICE-PRESIDENT AND PROGRAM CHAIR
Craig Smith
Saskatoon
Saskatchewan, Canada

SECRETARY-TREASURER (Canada)
Krystopher Chutko
Saskatoon
Saskatchewan, Canada

SECRETARY-TREASURER (USA)
Ken Rancourt
Conway
New Hampshire, USA

ESC PROCEEDINGS EDITORS

Krystopher Chutko
Saskatoon
Saskatchewan, Canada

Eli Deeb
Hanover
New Hampshire, USA

***PHYSICAL GEOGRAPHY* EDITORS**
Mauri Pelto
Dudley
Massachusetts, USA

STEERING COMMITTEE

Joan Ramage (Chair)
Bethlehem
Pennsylvania, USA

Sean Helfrich
Suitland
Maryland, USA

James Brylawski
Augusta
New Jersey, USA

David Robinson
New Brunswick
New Jersey, USA

Laura Thomson
Kingston
Ontario, Canada

Carrie Vuyovich
Hanover
New Hampshire, USA

THE PRESIDENT'S PAGE

My first Eastern Snow Conference was in 2007, hosted in St. John's, Newfoundland, and part of the first joint conference between the ESC, the Canadian Meteorological and Oceanographic Society (CMOS) and the Canadian Geophysical Union (CGU). Perhaps then it is fitting that I presided over the second such meeting in 2022. Of course, with the continued impacts of the COVID-19 pandemic, this meeting would be in an online/virtual format. Fortunately, with the lessons learned from the previous year of online conferences, we were able to create an informative, engaging, and entertaining meeting for more than 850 participants.

The scientific program included five sessions focused on snow and ice themes and were spread across three days. A total of 68 oral and poster presentations were offered encompassing a wide range of global cryosphere research. Oral presentations were conducted live over Zoom, while posters were presented using the Gather.Town app, a virtual social gathering where users could interact with others and discuss research in real time. I attempted to run the bar during the poster session, but unfortunately the app stops short of providing virtual drinks. Feedback on the virtual format was largely positive, though of course we all hoped to meet in person.

As usual, student research presentations were a highlight of the meeting; the future of snow and ice research is in good hands. I would like to send a special thanks to Sam Tuttle for organizing the daunting task of judging for our three student awards. The following students were recognized for their exemplary work:

- Megan Verfaillie, University of New Hampshire, awarded the Wiesnet Medal for best student presentation entitled "Fine-scale characterization of snowpack evolution using unpiloted aerial system lidar and SfM photogrammetry" (Wiesnet Medal and $750 cash prize)
- Alicia Pouw, Wilfred Laurier University, awarded the Campbell Scientific Canada Prize for "Evaluation of snow depth derived from ground penetrating radar on Canadian subarctic lakes" ($500 cash prize)
- Mariah Matias, Lehigh University, awarded the David Miller Award for best student poster entitled "Remote sensing of snowscapes and caribou (*Rangifer tarandus*) movement in the Northwest Territories of Canada" ($100 cash prize)

There are many people to thank for supporting my term as ESC President and the annual meeting. First and foremost is Craig Smith, ESC Vice-President, who performed double-duty as a CMOS organizer. The Local Arrangements Committee (led by Andrew Ireson and Yanping Li) and the Scientific Program Committee (led by Francis Zwiers) helped to make sure that the ESC was welcome. And a big thanks to all of the volunteers who made the meeting possible. I'd also like to acknowledge a pair of ESC Executive members who are moving on to bigger and better things: Ken Rancourt and Bart Forman. Both have played big roles behind the scenes to keep the ESC moving forward each year. Lastly, our thoughts are with ESC Executive Josh King and his family, who suffered a health emergency shortly after the meeting.

I will reiterate the comments of my predecessor here as well: the Eastern Snow Conference is committed to justice, equity, diversity, and inclusion. The ESC strives to be a welcoming venue for everyone and hopes to instill in the next generation of cryosphere researchers these important principles.

We are currently planning the 2023 ESC, the first to be held in person since 2019, in Easton, Pennsylvania, at the Nurture Nature Center. I hope to see everyone there!

Krystopher Chutko
78th President, Eastern Snow Conference

LIFE MEMBERS OF THE EASTERN SNOW CONFERENCE

The Eastern Snow Conference gratefully recognizes individuals who have made lifelong contributions to the study of snow and for their support of this organization.

JAMES FOSTER

BARRY GOODISON

GERRY JONES

JOHN METCALFE

HILDA SNELLING

DONALD WIESNET

In memoriam

PETER ADAMS

DON DUNLAP

ART ESCHNER

AUSTIN LOGAN

ROBERT B. SYKES

Wiesnet Medal

The Eastern Snow Conference encourages student research through its Wiesnet Medal. This medal is presented annually to the best student paper presented at the conference.

Year	Winner	Affiliation
2022	Megan Verfaillie	University of New Hampshire
2021	Daniela Krampe	Alfred-Wegener-Institut
2019	Zhibang Lv	University of Saskatchewan
2018	Jiyue Zhu	University of Michigan
2017	Caroline Dolant	Université de Sherbrooke
2016	Syed Mousavi	University of Michigan
2015	Nicolas Leroux	University of Saskatchewan
2014	Justin Hartnett	Syracuse University
2013	Andreas Dietz	Earth Observation Center/DFD, Germany
2012	Elizabeth Burakowski	University of New Hampshire
2011	Kathryn Semmens	Lehigh University
2010	Simon von de Wall	University of Victoria
2009	Si Chen	Dartmouth College
2008	Chris Fuhrmann	University of North Carolina at Chapel Hill
2007	*Not awarded*	
2006	Y.C. Chung	University of Michigan
2005	M. Javan-Mashmool	Université du Québec à Chicoutimi
2004	J. Farzaneh-Dehkordi	Université du Québec à Chicoutimi
2003	Alexandre Langlois	Université de Sherbrooke
2002	Patrick Ménard	Université de Laval
2001	C. Tavakoli	Université du Québec à Chicoutimi
2000	*Not awarded*	
1999	S. Brettschnieder	Université du Québec à Chicoutimi
1998	Andrew Grundstein	University of Delaware
1997	Newell Hedstrom	University of Saskatchewan
1996	Suzanne Hartley	University of Denver
1995	Paul Wolfe	Wilfred Laurier University
1994	G.E. Mann	University of Michigan
1993	G. Devarennes	Université du Québec à Québec
1992	D.W. Cline	University of Colorado
1991	D. Samelson	Cornell University
1990	A.K. Abdel-Zaher	University of New Brunswick
1989	A. Giguere	McGill University
1988	Mauri Pelto	University of Maine
1987	Cameron Wake	Wilfred Laurier University
1986	Craig Allan	Trent University
1985	Robert Speck	Rensselaer Polytechnic Institute
1984	N.K. Kalliomaki	Laurentian University
1983	David Beresford	Trent University
1982	*Not awarded*	
1981	Jeffrey Patch	University of New Brunswick
1980	Bryan Wolfe	Trent University
1979	Margaret Leech	McGill University
1978	Michael English	Trent University
1977	Don McLaughlin & George Duggan	Rensselaer Polytechnic Institute
1976	Dwayne McMurter	Trent University
1975	Nigel Allan	Syracuse University
1974	*Not awarded*	
1973	Stan Mathewson	Trent University

David Miller Award

The David Miller Award is presented to the best student poster at the annual Conference.

Year	Winner	Affiliation
2022	Mariah Matias	Lehigh University
2021	Madison Woodley	Syracuse University
2019	Gaohong Yin	University of Maryland
2018	Justin Pflug	University of Washington
2017	Elizabeth Ryan	University of Maryland
2016	Oliver Wigmore	The Ohio State University
2015	Charles Papasodoro	Université de Sherbrooke
2014	Laura Thomson	University of Ottawa
2013	Nastaran Saberi	University of Waterloo
2012	Johnathan Sugg	Appalachian State University
2011	*Not awarded*	
2010	Nicolas Svacina	University of Waterloo
2009	Iliyana Dobreva	Texas A&M University
2008	Christopher Karmosky	Pennsylvania State University

Campbell Scientific Canada Prize

The Campbell Scientific Canada prize is awarded to student research showing the most innovative use of technology in the gathering of data.

Year	Winner	Affiliation
2022	Alicia Pouw	Wilfred Laurier University
2021	Fraser King	University of Waterloo
2019	Alex Mavrovic	Université de Sherbrooke
2018	*Not awarded*	
2017	Abigail Dalton	University of Ottawa
2016	*Not awarded*	
2015	*Not awarded*	
2014	Aaron Campbell	University of Waterloo
2013	*Not awarded*	
2012	Andrew Kasurak	University of Waterloo
2011	Mathieu Beaulieu	University of British Columbia
2010	*Not awarded*	
2009	*Not awarded*	
2008	Nicholas Kinar	University of Saskatchewan
2007	Gro Lilbaek	University of Saskatchewan
2006	Gernot Koboltschnig	University of Natural Resources and Applied Life Sciences, Vienna
2005	Warren Helgason	University of Saskatchewan

Session #1: ESC General Session

Conductive Heat Flow through Temperate Region Lake Ice

LAURA ALVAREZ SALINAS[1] AND LAURA BROWN[1]

ABSTRACT

Understanding the conductive heat flow through lake ice is important to understanding the future of lake ice in a changing climate. Temperate lake ice has a different composition compared to arctic lake ice due to the warmer and wetter climate experienced in more southern latitudes. With limited research, there is a need to study how conductive heat flow travels through the ice layers in temperate regions. Consequently, this research focuses on expanding research in temperate lake ice by providing the first in-depth look at heat flux through the predominantly white ice layers found in the temperate region. This research investigates the temperature gradients within both black and white ice, and the snowpack on top of the ice, while also looking at lake ice thickness over the span of six years from 2017 to 2022. The area of study consists of two lakes in Central Ontario: Macdonald Lake and Clear Lake. Snow and ice measurements have been collected along 4 sampling transects (snow depth, density, temperature, grain type, albedo, ice thickness, ice type) while heat flow was measured through the ice using thermistors at a single location. Data collected has shown climate variability throughout the collection period, with thicker white ice in the most recent years compared to earlier years. Moreover, the data collected from the thermistors allows an insight on how conductive heat flow occurs through both the black and white ice. Furthermore, snowpack data also reflects climate variability by exhibiting different patterns in the temperature profile which has reflected variations in degrees of annual slushing. This research aids to show how heat transfers through the ice layers and allow for a more thorough understanding of the effect that heat flow has on temperate region lake ice, greatly expanding the field.

[1] Department of Geography, Geomatics and Environment, University of Toronto Mississauga, Mississauga, ON, Canada

Monitoring Freshwater Lake Ice Thickness and Ice Bottom Roughness in Central Ontario

NOAH BACAL[1] AND LAURA BROWN[1]

ABSTRACT

Lake ice thickness is an important parameter for understanding hydrological changes and implications in the Northern Hemisphere, yet there is a stark lack of ground-based lake ice data across Canada. Remote sensing techniques are increasingly used to address this void. Satellite-based ice thickness retrievals are becoming more reliable and widely used for Arctic sea ice, however, these methods are not yet viable for most freshwater lakes. Currently, satellite-based lake ice thickness can only be retrieved for very large lakes or small lakes where the ice freezes to the lakebed. Current approaches of lake ice thickness retrieval on other lakes, including those in the mid-latitudes, focus on RADAR backscatter analysis and are working towards accounting for the distortion of the backscatter signal at the ice-water interface, caused by roughness on the underside of the ice layer. To date, there is no field-based method to quantify the roughness of the ice bottom surface in deeper freshwater lakes. Using *in situ* measurements of lake ice thickness, and by measuring roughness from physical imprints of the ice bottom surface, this project aims to identify a correlation between these two characteristics that will better inform how ice bottom roughness impacts total ice thickness on mid-latitude freshwater lakes. Preliminary results from measurements of early-season roughness indicate ice bottom features ranging from 0.010 mm to 2.071 mm in height protruding from the ice bottom surface. Measurements of the total imprint depth per unit area of the ice bottom surface range from 0.198 mm cm^{-2} to 1.323 mm cm^{-2}. Further sampling of ice bottom roughness characteristics is planned to better relate observed roughness characteristics with airborne ice thickness retrievals using Ku-band and L-band RADAR backscatter analysis, towards developing space-based ice thickness retrievals in the future.

[1] Department of Geography, Geomatics and Environment, University of Toronto Mississauga, Mississauga, ON, Canada

Observationally Constraining a Snow-on-Sea-Ice Model to Estimate Arctic Snow and Sea Ice Thickness with Associated Uncertainties

ALEX CABAJ[1], PAUL KUSHNER[1], AND ALEK PETTY[2]

ABSTRACT

Snow on Arctic sea ice plays many, sometimes contrasting roles in Arctic climate feedbacks. Snow depth is also a key input for sea ice thickness derived from ice altimetry measurements, such as satellite lidar observations from ICESat-2. Making direct snow measurements is logistically challenging in the Arctic due to the remoteness of the region, so basin-wide snow depth on sea ice is difficult to observationally constrain. Likewise, uncertainties in snow-on-sea-ice and in derived sea ice thickness are challenging to quantify.

Snow-on-sea-ice models, such as the NASA Eulerian Snow-On-Sea-Ice Model (NESOSIM), can provide basin-wide snow depth and density estimates over Arctic sea ice. NESOSIM includes free parameters which dictate the strength of snow densification and loss processes, but these parameters are not directly observationally well-constrained. We present a calibration of NESOSIM to snow depth and density observations using a Metropolis Markov Chain Monte Carlo method. This method provides estimates of the model free parameters and their uncertainty distributions. We propagate the parameter uncertainty estimates through NESOSIM to produce uncertainty estimates for model-output snow depth and density. Finally, we estimate the resulting sea ice thickness using NESOSIM snow output and ICESat-2 altimetry measurements, and quantify the contribution of snow uncertainty to uncertainty in sea ice thickness.

[1] Department of Physics, University of Toronto, ON, Canada
[2] NASA Goddard Space Flight Center, Greenbelt, MD, USA

Evaluating Fourteen Gridded SWE Products using Airborne Gamma Radiation SWE and Snow Course Transects

EUNSANG CHO[1], COLLEEN MORTIMER[2], LAWRENCE MUDRYK[2], CHRIS DERKSEN[2], CARRIE VUYOVICH[1], AND MIKE BRADY[2]

ABSTRACT

Snow-related research in hydrological and climate sciences at continental and global scale typically rely on spatially distributed snow water equivalent (SWE) products. Recently, many gridded SWE datasets have been developed from climate, snow, hydrological sciences community, which must be validated with independent and reliable observations in various environments. Such validation is challenging due to limited spatial coverage of available reference data (e.g., snow course transects, and airborne flight lines) particularly in complex terrains, thus validation with multiple sources of reference data is required. In this study, we quantify 1) the spatial representativeness of SWE estimates from airborne gamma radiation measurements through comparisons at different distances from snow course transects and 2) the performance of fourteen gridded SWE products using gamma and snow course reference datasets over the continental U.S. and southern Canada.

Results show strong agreement between the gamma measurements and snow course transects at distances up to at least 50 km in the northeastern U.S. However, there is limited agreement beyond 5 km in the western U.S. likely due to more limited spatial representativeness of the gamma observations in complex topography. Among fourteen gridded SWE products, ERA5-Land and University of Arizona SWE showed the best performance with both reference SWE observations. This work supports global SWE evaluation efforts within the European Space Agency (ESA) Satellite Snow Product Intercomparison and Evaluation Exercise (SnowPEx), by providing guidance on integrating different sources of reference data.

[1] NASA Goddard Space Flight Center, Greenbelt, MD, USA
[2] Environment and Climate Change Canada, Toronto, ON, Canada

How do Microwaves Interact with Freshwater Ice, and Why is it so Important?

GRANT GUNN[1] AND JAKE FERGUSON[1]

ABSTRACT

We present the history of literature's understanding of microwave interactions with freshwater ice, and the potential that the current generation of spaceborne SAR sensors present for ice property retrieval. In the 1970s, Side Looking Airborne Radar (SLAR) imaged frozen thermokarst lakes on the coast of the Alaskan north coast as a by-product of observing sea ice features. Elongated lakes were bright in the imagery compared to the dark tundra, which was hypothesized to be the result of long cylindrical bubble inclusions within the ice column causing a double bounce of the incident signal (scattered by the bubble, reflected off the ice-water interface and back to the sensor).

The double-bounce hypothesis was pervasive into the 2010s, until fully polarimetric acquisitions (quad-pol, HH/VV/VH/HV) became accessible over northern lakes. Quad-pol SAR data allowed for the application of polarimetric decomposition algorithms (e.g. Freeman-Durden) which provide a measure of the relative contribution of surface, double-bounce, and volume scatter interactions with targets based on the covariance matrix. Recent publications and the results presented in this study indicate that lakes, regardless of frequency (L- to X-band), incidence angle, ice structure, or morphometry overwhelmingly display odd-bounce surface interactions from the ice-water interface as the dominant scattering mechanism (as opposed to double-bounce). With the understanding that the dominant source of scatter occurs at the ice-water interface, the direct retrieval of changes in ice thickness highlights the potential ice properties that can be obtained using the current generation of SAR satellite constellations. This study presents preliminary analysis of ice thicknesses directly measured from TanDEM-X using repeat-pass interferometry for multiple regions in northern Canada and Alaska.

[1] Department of Geography and Environmental Management, University of Waterloo, Waterloo, ON, Canada

Comparison of VIIRS and NOAA/Rutgers IMS Snow-Cover Frequency Maps of North America

DOROTHY HALL[1], GEORGE RIGGS[2], NICOLO DIGIROLAMO[2], THOMAS W. ESTILOW[3], AND DAVID ROBINSON[3]

ABSTRACT

Though the mean winter maximum snow-cover extent (SCE) can cover up to ~50 million km^2 over Northern Hemisphere land areas, there is a large amount of variability in the month-to-month and interannual SCE in the Northern Hemisphere where ~98 percent of Earth's seasonal snow is located. We use the Suomi National Polar-orbiting Partnership (S-NPP) Visible Infrared Imaging Radiometer Suite (VIIRS) daily cloud-gap filled 375-m resolution Normalized-Difference Snow Index (NDSI) maps derived from the VNP10A1F NASA standard data snow-cover product to develop a time series of SCE in North America. To assess the data-product quality, we compare VIIRS snow-mapping results with a climate data record-quality SCE product generated at the Rutgers University Global Snow Lab from the NOAA/U.S. National Ice Center 24-km Interactive Multisensor Snow and Ice Mapping System (IMS) (the NOAA/Rutgers IMS product). To compare with the NOAA/Rutgers IMS snow maps, we developed monthly VIIRS maps showing percent of days in a month that were snow covered for the first six months of 2019, degrading the resolution of the VIIRS maps to 24 km so that the VIIRS and NOAA/Rutgers IMS maps could be compared directly on the same grid. Results show good correspondence between the VIIRS and NOAA/Rutgers IMS monthly SCE snow-frequency maps for North America, with 88.2 – 99.8 percent agreement for total snow cover. The correspondence between the two independent datasets increases our confidence in satellite-derived SCE from both products.

INTRODUCTION

About 98 percent of Earth's seasonal snow-cover extent (SCE) is found in the Northern Hemisphere, with a mean winter maximum SCE over land areas that can cover ~50 million km^2 (Sturm, 2020). Seasonal snow cover is a critical component of Earth's energy balance, in part, because it reflects much of the incoming solar radiation back to space. It is also an excellent insulator of the ground beneath. Continental snowmelt contributes to spring and summer streamflow and is an important water resource for billions of people worldwide as well as being a hazard that can cause flooding and other environmental problems. In forested areas, snowmelt recharges aquifers which can be important sources of municipal water supply (Malle *et al.*, 2019; Swistock, 2020). Changes in seasonal snow cover have a profound effect on Earth's energy balance and have substantial societal consequences, some of which might be positive, but most of which are problematic. The only way to monitor continental-scale SCE on a daily or sub-daily basis is via satellite remote sensing.

---

[1] Earth System Science Interdisciplinary Center, University of Maryland, College Park, MD, USA

[2] SSAI / NASA Goddard Space Flight Center, Greenbelt, MD, USA

[3] Department of Geography, Rutgers University, Piscataway, NJ, USA

To evaluate the monthly extent and variability of SCE in North America, we provide comparisons of the Suomi National Polar-orbiting Partnership (S-NPP) Visible Infrared Imaging Radiometer Suite (VIIRS) daily cloud-gap filled standard 375-m resolution snow-cover maps, VNP10A1F (Riggs and Hall, 2020; Riggs and Hall, 2021) with the monthly SCE maps developed by the Rutgers University Global Snow Lab (RUGSL) from the NOAA/US National Ice Center 24-km Interactive Multisensor Snow and Ice Mapping System (IMS) (the NOAA/Rutgers IMS product). The NOAA/Rutgers IMS maps are quality-controlled for consistency, with aggregation to monthly maps and SCE area calculations performed according to the Rutgers routine (Robinson *et al.*, 1993; Estilow *et al.*, 2015). NASA VIIRS SCE data products will extend the MODerate-resolution Imaging Spectroradiometer (MODIS) snow cover Environmental Science Data Record that began in 2000 and may end in 2023 when plans call for the MODIS instruments to be decommissioned [*https://www.earthdata. nasa.gov/learn/articles/from-terra-to-terra-firma*]. The VIIRS SCE maps permit detailed studies of snow cover at the watershed scale, as well as multi-year trends in SCE at continental and basin scales to be calculated when combined with the longer MODIS record (Hall *et al.*, 2019). VIIRS SCE products will likely be available through the 2030s on the Joint Polar Satellite System (JPSS) (Justice *et al.*, 2013) and will thus be available to extend the MODIS SCE record and potentially contribute to a moderate-resolution SCE climate data record when 30 years of MODIS and VIIRS data are available.

Absolute validation of the VIIRS SCE maps is not possible because there is no 'gold-standard,' validated hemispheric or global snow-cover dataset available. Thus, in this work, we begin to compare the VIIRS SCE with the independently-developed NOAA/Rutgers IMS SCE time series.

BACKGROUND

On 28 October 2011 the first in a series of VIIRS instruments was launched on the S-NPP satellite. VIIRS permits the generation of daily SCE maps at 375-m resolution (Figure 1) which is an improvement on, but compatible with, the 500-m spatial resolution of the MODIS snow-cover maps (Riggs *et al.*, 2017; Riggs and Hall, 2020) and permits detailed studies of snow cover at the watershed scale globally. A complete record of daily, global VIIRS snow-cover products (VNP10A1F) from 19 January 2012 to the present is available from the NASA DAAC at the National Snow and Ice Data Center (NSIDC). VNP10A1F is a cloud-gap-filled (CGF) snow map product developed from the VNP10A1 SCE standard NASA product (Riggs and Hall, 2021). The fully automated algorithm to produce the VIIRS CGF products is modeled after, and compatible with the MODIS CGF algorithm (Hall *et al.*, 2019; Riggs *et al.*, 2019; Riggs *et al.*, 2020; Riggs and Hall, 2021). Numerous authors have used MODIS snow-cover products to show interannual change in continental- and basin-scale SCE worldwide (for example, see: Parajka and Blöschl, 2008; Wang *et al.*, 2017; Aguirre *et al.*, 2018; Hammond *et al.*, 2018; Saavedra *et al.*, 2018; Choudhury *et al.*, 2021; Hall *et al.*, 2021).

VNP10A1F uses a Normalized Difference Snow Index (NDSI) data layer, providing snow cover for each pixel or cell, in increments ranging from NDSI = 0 to 100 (Figure 1). Lower NDSI values are associated with lower snow-cover fractions and NDSI = 100 is associated with a pixel or cell that is completely snow covered (Riggs *et al.*, 2021). The resulting maps provide global, daily, cloud-free SCE for each pixel or cell. There are uncertainties in the VIIRS CGF snow-cover products due to the: 1) original VNP10A1 products, and 2) CGF algorithm. These uncertainties are discussed in the literature (e.g., see Hall *et al.*, 2019) and in the MODIS and VIIRS user guides (Riggs *et al.*, 2018; Riggs and Hall, 2021) and need not be repeated here.

The NOAA/Rutgers IMS SCE maps are derived from daily analyses of mostly-visible imagery from a constellation of geostationary and polar-orbiting satellites and sensors, and ground station data in addition to VIIRS band data. The maps are generated by trained meteorologists, thus are not fully automated (Helfrich *et al.*, 2018). At Rutgers, data are subject to quality control procedures established over decades of working with IMS products and their predecessors, that includes establishing consistent land masks, grid-cell areas, and approaches to generating a variety of daily, weekly, seasonal, and annual products (Robinson *et al.*, 1993; Estilow and Robinson, 2015). Daily IMS SCE data consist of binary (snow/no snow) values for each 24-km grid-cell, which are used to calculate monthly SCE frequency.

RESULTS: COMPARISON WITH THE 24-KM RUGSL IMS SNOW-COVER EXTENT SNOW-FREQUENCY MAPS

Though there is no "gold standard" global snow-cover map with which to compare the VIIRS time series of SCE, valuable information can be gained from comparing independently-derived SCE maps. Here we compare VIIRS and NOAA/Rutgers IMS snow-frequency maps for North America.

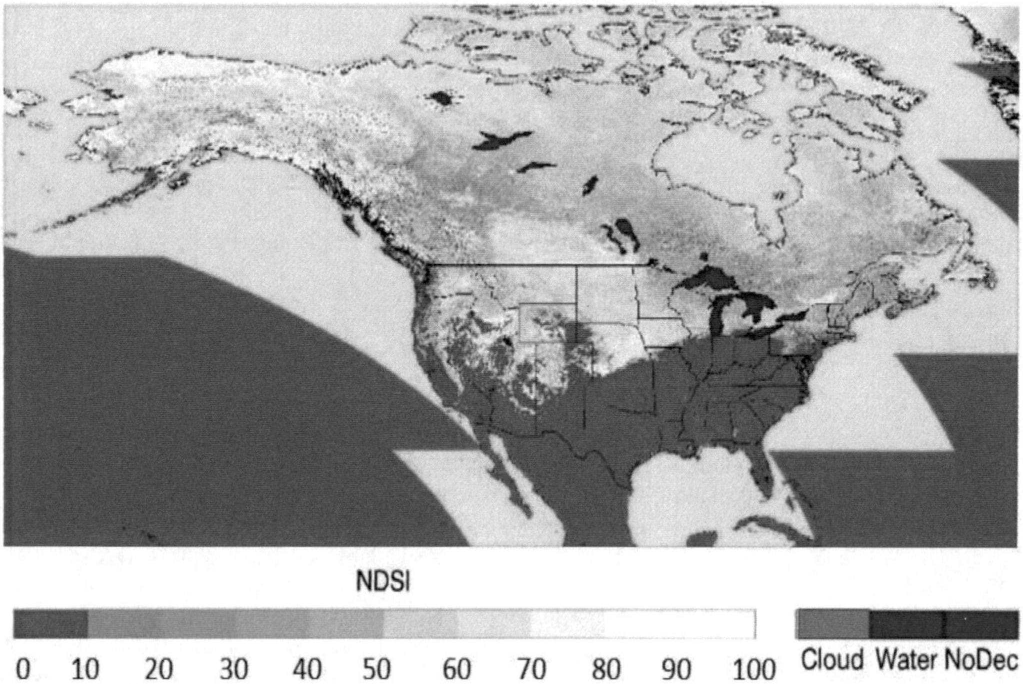

Figure 1. VNP10A1F VIIRS normalized-difference snow index (NDSI) snow-cover extent (SCE) map of North America, 1 March 2019.

To enable the comparison, the VIIRS and NOAA/Rutgers IMS maps were projected to a common grid with a spatial resolution of 24 km^2. A binary map was produced from the VNP10A1F map product using a value of NDSI = 10 or greater to indicate snow in a cell, to compare with the binary NOAA/Rutgers IMS map. For each month, the daily VNP10A1F tiles were used to produce tiles indicating the percent of days with snow cover for each 375-m cell. The North America tiles were digitally stitched together to produce a 375-m map of North America, from which the SCE was determined.

We compared monthly VIIRS and NOAA/Rutgers IMS SCE snow-frequency maps for the first six months of 2019 (Table 1). Results show excellent agreement overall, but also some differences that we attribute to differences in methodologies used to develop the maps, sensor differences, differences in spatial resolution and the handling of cloud-cover.

Table 1. Comparison of total SCE mapped using the NOAA/Rutgers IMS and VIIRS SCE maps for the first six months of 2019, in km^2.

Month in Year 2019	NOAA/Rutgers (km^2)	VIIRS (km^2)	Percent agreement
January	15,860,500	15,749,400	99.7
February	16,242,000	16,281,600	99.8
March	15,105,800	15,246,900	99.1
April	11,485,600	11,715,400	98.0
May	7,861,700	7,971,700	98.6
June	3,832,500	4,315,100	88.2

Figure 2, below, shows a comparison of March 2019 snow maps of North America. The total amount of snow cover in the NOAA/Rutgers IMS map (left) was 15,105,800 km^2 *vs.* 15,246,900 km^2, for the VIIRS map (right).

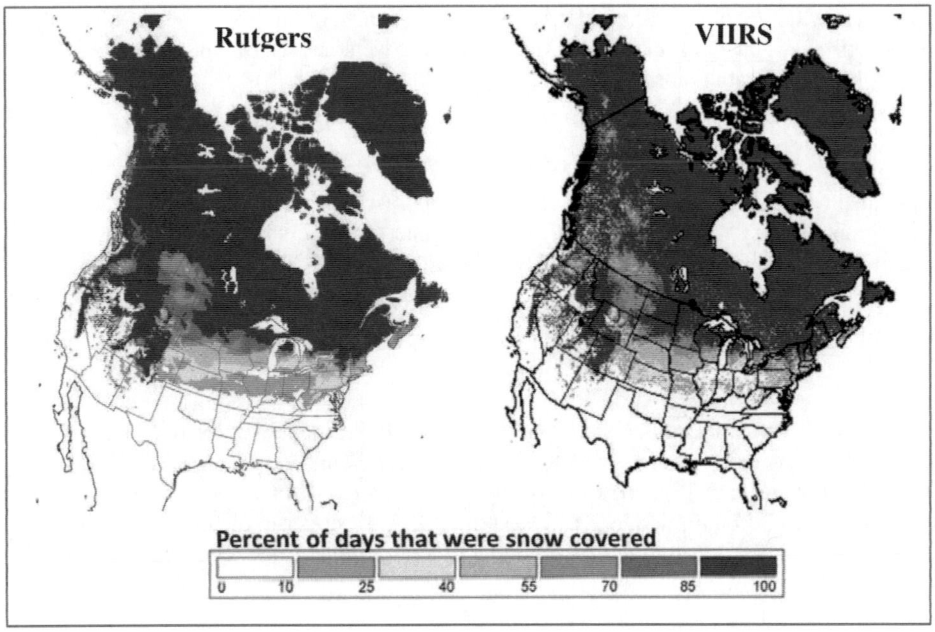

Figure 2 (a & b). Comparison of the NOAA/Rutgers IMS monthly SCE map (Left) from March 2019 with a VIIRS monthly SCE map (Right) from March 2019, both produced to provide percent of days that were snow covered.

The algorithms and methodologies to produce these maps are entirely different. As discussed previously, the NASA VIIRS maps are derived using only VIIRS band data and the VIIRS cloud mask product to produce a fully automated SCE map product. The NOAA/Rutgers IMS SCE maps are generated using multiple satellites with primarily visible sensors, and ground station data in addition to VIIRS band data and is not fully automated. Though there are many similarities in the maps, hemispheric-scale SCE maps developed using different input data and at different spatial resolutions are expected to provide some differences in SCE. For one thing, the maps may be derived from imagery and ground station data acquired at different times of the same day which can lead to some differences especially in areas having shallow or ephemeral snow. Additionally, differences in cloud screening methodologies between the VIIRS and NOAA/Rutgers IMS SCE mapping techniques influence the amount of snow that is mapped.

DISCUSSION AND CONCLUSION

There are important differences between the NOAA/Rutgers IMS and NASA VIIRS SCE maps, yet when monthly maps of percent of days that are snow covered are gridded identically, there is excellent overall agreement between the maps regarding the total amount of SCE. The NOAA/Rutgers IMS maps have the advantage of human intervention in their production and both satellite and *in situ* measurements are used to produce the maps. Thus, the use of a cloud mask is not needed. On the other hand, the VNP10A1F map product is derived from the VNP10A1 product that includes cloud cover, then the cloud cover is masked using the VIIRS cloud mask product. The VIIRS maps used in this comparative study are developed from the cloud-gap filled NASA VIIRS standard product, VNP10A1F. The VNP10A1F algorithm looks back on previous cloud-free days to map snow when the current day is cloudy. Full-resolution VIIRS SCE maps at 375-m spatial resolution provide detail that is not possible in the IMS 24-km resolution SCE maps. Despite the differences, the agreement achieved when mapping snow in North America, at least for the first six months of 2019, is encouraging.

In future work, we will study a longer time period and focus on areas of disagreement between the two SCE products. We will evaluate specific differences between them in concert with other information such as MODIS and VIIRS single-band data and Landsat-8 and -9 higher resolution (30 m) data. This will assist in determining the strengths and weaknesses of both the VIIRS and NOAA/Rutgers IMS SCE data records.

Acknowledgements. The authors thank Walt Clark / NOAA/U.S. National Ice Center for discussions related to the IMS snow product. Some of this research was supported by the NASA Earth Science, grant number 80NSSC18K1690, and by a grant to Rutgers from the NOAA National Centers for Environmental Information (NOAA/NCEI) through Riverside Technology, Inc.

REFERENCES

Aguirre F, Carrasco J, Sauter T, Schneider C, Gaete K, Garín E, Adaros R, Butorovic N, Jaña R, Casassa G. 2018. Snow cover change as a climate indicator in Brunswick Peninsula, Patagonia. *Frontiers in Earth Science*, **6**: 130. *https://doi.org/10.3389/ feart.2018.00130*

Choudhury A, Yadav AC, Bonafoni, SA. 2021. Response of snow cover to the climate in the northwest Himalaya (NWH) using satellite products. *Remote Sensing*, **13**: 655. *https:// doi.org/ 10.3390/rs13040655*

Estilow TW, Young AH, Robinson DA. 2015. A long-term Northern Hemisphere snow cover extent data record for climate studies and monitoring. *Earth System Science Data*, **7**(1): 137-142.

Hall DK, Riggs GA, DiGirolamo NE, Román MO. 2019. Evaluation of MODIS and VIIRS cloud-gap-filled snow-cover products for production of an Earth science data record. *Hydrology and Earth System Sciences*, **23**: 5227-5241. *https://doi.org/10.5194/hess-23-5227-2019*

Hall DK, O'Leary III D, DiGirolamo NE, Miller W, Kang DK. 2021. The role of declining snow cover in the desiccation of the Great Salt Lake, Utah, using MODIS data. *Remote Sensing of Environment*, Terra@20 Special Issue **252**: 112106. *https://doi.org/10.1016/ j.rse.2020.112106*

Hammond JC, Saavedra FA, Kampf SK. 2018. Global snow zone maps and trends in snow persistence 2001-2016. *International Journal of Climate*, **38**(12): 4369–4383. *https://doi:10. 1002/joc.5674*

Helfrich S, Ki M, Kongoli C, Nagdimunov L, Rodriguez E. 2018. Interactive Multisensor Snow and Ice Mapping System Version 3 (IMS V3), IMS Algorithm Theoretical Basis Document Version 2.5, NOAA NESDIS STAR. *https://nsidc.org/drupal/interactive-multisensor-snow-and-ice-mapping-system-version-3-ims-v3-algorithm-theoretical*

Justice CO, Román MO, Csiszar I, Vermote EF, Wolfe RE, Hook SJ, Friedl M, Wang Z, Schaaf CB, Miura R, Tschudi M, Riggs G, Hall DK, Lyapustin AI, Davidson C, Masuoka EJ. 2013.

Land and cryosphere products from Suomi NPP VIIRS: Overview and status, *Journal of Geophysical Research – Atmospheres*, **118**(17): 9753-9765.

Malle J, Rutter N, Mazzotti G, Jonas T. 2019. Shading by trees and fractional snow cover control the subcanopy radiation budget. *Journal of Geophysical Research – Atmospheres*, **124**(6): 3195-3207. *https://doi:10.1029/2018JD029908*

Parajka J, Blöschl G. 2008. Spatio-temporal combination of MODIS images–potential for snow cover mapping. *Water Resources Research*, **44**(3): W03406. *https://doi.org/10.1029/2007WR006204*

Riggs GA, Hall DK. 2020. Continuity of MODIS and VIIRS snow cover extent data products for development of an Earth Science Data Record. *Remote Sensing*, **12**: 3781. *https://doi:10.3390/rs12223781*

Riggs GA, Hall DK. 2021. VIIRS snow products user guide for Collection 1 (C1).

Riggs GA, Hall DK, Román MO. 2017. Overview of NASA's MODIS and Visible Infrared Imaging Radiometer Suite (VIIRS) snow-cover. *Earth System Science Data*, **9**: 1–13. *https://doi.org/10.5194/essd-9-765-2017*

Riggs GA, Hall DK, Román MO. 2018. MODIS Snow Products Collection 6.1 User Guide, Version 1.0 [*https://modis-snow-ice.gsfc.nasa.gov/?c=userguides*][*http://modis-snow-ice.gsfc.nasa.gov/?c=userguides*]

Riggs GA, Hall DK, Román MO. 2019. *VIIRS/NPP Snow Cover Daily L3 Global 375m SIN Grid, Version 1*. Boulder, Colorado USA. NASA National Snow and Ice Data Center Distributed Active Archive Center. *https://doi.org/10.5067/VIIRS/VNP10A1.001*

Robinson DA, Dewey F, Heim Jr. R. 1993. Global Snow Cover Monitoring: An Update. *Bulletin of the American Meteorological Society*, **74**: 1689–1696. *https://doi:10.1175/1520-0477(1993)074<1689:GSCMAU>2.0.CO;2*

Rutgers University Global Snow Lab. Available online: [*https://climate.rutgers.edu/snowcover/*] (accessed on 10 June 2021).

Saavedra FA, Kampf SK, Fassnacht SR, Sibold JS. 2018. Changes in Andes snow cover from MODIS data, 2000–2016. *The Cryosphere*, **12**(3): 1027-1046. *https://doi.org/10.5194/tc-12-1027-2018.*

Sturm M. 2020. *A Field Guide to Snow*. University of Alaska Press.

Swistock B. 2020. The importance of snow and forests to Pennsylvania streams and groundwater, Penn State Extension Service. *https://extension.psu.edu/the-importance-of-snow-and-forests-to-pennsylvania-streams-and-groundwater*

Wang X, Wu C, Wang H, Gonsamo A, Liu Z. 2017. No evidence of widespread decline of snow cover on the Tibetan Plateau over 2000–2015. *Scientific Reports*, **7**(1): 1-10.

Development of Snow Wetness *in situ* Data Distribution Survey

VICKI JAGDEO[1], MAHBOUBEH BOUESHAGH[1], JOAN RAMAGE[1], MARY BRODZIK[2], AND ELIAS DEEB[3]

ABSTRACT

Snow wetness is the amount of liquid water present in a snowpack and is an important physical parameter of snow. Snowmelt timing is vital for water storage and resource management, and possible environmental hazards such as flooding. However, snowmelt timing is ephemeral, spatially variable, and difficult to estimate because of the inaccessibility and lack of field observations in many regions. *In situ* measurements of snow wetness are critical for documenting snow hydrology, including seasonal snow changes that affect runoff timing and discharge. Understanding these changes has important implications in regional climate variability and vehicle mobility. Therefore, ground-based observations are essential for researchers and scientists to develop improved remote sensing and modeling approaches to snowmelt processes in remote and vast areas. We are developing a community survey to collect information about existing ground-based snow wetness data. The survey will serve to organize information about the data, like measurement type, technology and tools used, date and location, altitude, elevation, frequency, data accessibility, and references. We will organize the information by region, communicate additional needs from individuals, and make the collection available to others. Once collected, the compiled responses will be shared with the snow community. We hope for broad community engagement and a range of dataset date, location, and scope. This work will address the need for collecting and sharing existing ground-based data on snow wetness and assess data gaps. Since existing data sets include variable intervals with different collection methods and locations, this survey will benefit the community to have a searchable list of existing datasets and sampling approaches. Contributors will retain their own datasets. The development of this resource provides a collective space for community knowledge of existing *in situ* observations and their application in snow science.

[1] Department of Earth and Environmental Sciences, Lehigh University, Bethlehem, PA, USA

[2] Cooperative Institute for Research in Environmental Sciences, University of Colorado Boulder, Boulder, CO, USA

[3] US Army Engineer Research and Development Center, Cold Regions Research and Engineering Laboratory, Hanover, NH, USA

DeepPrecip: A Deep Neural Network for Retrievals of Precipitation

FRASER KING[1], GEORGE DUFFY[2], LISA MILANI[3], CHRISTOPHER FLETCHER[1], CLAIRE PETTERSEN[4], AND KERSTIN EBELL[5]

ABSTRACT

Remotely-sensed vertical radar reflectivity profiles of the lower atmosphere can be linked to precipitation through empirical power-law relationships at high spatiotemporal frequency. These relationships are tightly coupled to particle phase, shape, size and density, which contributes to uncertainty and error when applied to unseen data. In this work, we develop an alternative precipitation retrieval using a deep convolutional multilayer perceptron (DeepPrecip), to estimate surface precipitation from eight measurement sites across the northern hemisphere. Using a combination of K-band micro-rain radar (MRR) retrievals up to 3 km, surface meteorology observations, and ERA-5 atmospheric data, a total of N = (60307 x 45) data points spanning 2012 to 2020 are used to train DeepPrecip against collocated 20-minute-average *in situ* precipitation accumulation records. DeepPrecip displays strong predictive skill on unseen data and is able to accurately model snowfall with a mean square error (MSE) of 1.3×10^{-4} mm, rainfall with a MSE of 1.2×10^{-4} mm, and displays positive Pearson correlation of $r = 0.64$. These MSE values are approximately 170% lower, on average, than current commonly used methods. We find that both near-surface and top-of-profile radar observations provide the most important information contributing to model skill. These results highlight the value of the relative structure of the full vertical column in contributing to improved retrieval accuracy. DeepPrecip also displays skill in capturing changes in precipitation magnitude across different sites/climates, and over multiple seasons; without explicit descriptions of precipitating particle microphysics or geospatial covariates being provided to the model. This research reveals the important role for deep learning in extracting predictive information about precipitation from radar retrievals.

[1] Department of Geography and Environmental Management, University of Waterloo, Waterloo, ON, Canada
[2] NASA Jet Propulsion Laboratory, Pasadena, CA, USA
[3] NASA Goddard Space Flight Center, Greenbelt, MD, USA
[4] Climate and Space Science and Engineering, University of Michigan, Ann Arbor, USA
5 Institute of Geophysics and Meteorology, University of Cologne, Cologne, Germany

Snow Depth on Sea Ice Record from 1955-2019 in the Canadian Arctic and Development Plans for Multi-Satellite Snow Depth Retrieval

HOI MING LAM[1], TORSTEN GELDSETZER[1], STEPHEN HOWELL[2], AND JOHN YACKEL[1]

ABSTRACT

Using *in situ* snow depth measurements at 11 Canadian study sites from the period 1955-2019, we study the intra-annual and decadal trends in snow accumulation on landfast sea ice and on terrestrial ground within the Canadian Arctic. Ice chart data acquired via the Canadian Ice Service are used to establish sea ice break-up and freeze-up dates and assess their impact on snow depth evolution. We find that on-ice and on-land snow accumulation in autumn differ due to the lag between the freeze-up and the first snow of the season. Once sea ice consolidates, on-ice and on-land snow depth become positively correlated in winter ($p < 0.05$). The mean seasonal rate of snow accumulation on sea ice from September to April is 3.2 ± 0.6 cm per month across the Canadian Arctic. Snow depth on terrestrial land is generally higher than on sea ice in the southern Canadian Arctic by up to 20 to 30 cm; but snow depth on sea ice tends to exceed that on land in the northern Canadian Arctic from winter to spring. Results have been submitted for publication in a peer-reviewed journal (in print).

As a follow-up study, a machine learning (ML) algorithm is being developed to estimate winter snow depth on landfast sea ice in the Canadian Arctic using measurements from satellite altimeters (ICESat-2 and CryoSat-2) and radar scatterometers (Sentinel-1 and RADARSAT-CM), under the first-order assumption that rougher sea ice, which is signified by higher SAR backscatter, would entrap thicker snow. Preliminary results indicate some correspondence between the altimetric surface heights and SAR backscatter coefficient (sigma-nought). Our proposed approach can provide greater spatial and temporal coverages for snow depth estimates on Arctic landfast sea ice. A field campaign is planned in April 2022 to sample snow on sea ice at near-coincident CRYO2ICE tracks to provide validation datasets.

[1] Department of Geography, University of Calgary, Calgary, AB, Canada
2 Environment and Climate Change Canada, Toronto, ON, Canada

Remote Sensing of Snowscapes and Caribou (*Rangifer tarandus*) Movement in the Northwest Territories of Canada

MARIAH MATIAS[1], JOAN RAMAGE[1], MARY BRODZIK[2], AND KRISTEN HEROY[3]

ABSTRACT

Caribou play a crucial role in nutrient cycling and turnover, and they are an essential herbivore and prey animal in northern ecosystems. This keystone species undergoes exceptionally large, annual synchronized migrations on the order of thousands of kilometers, triggered by shared environmental stimuli. Relationships between climate factors and caribou populations and their migration patterns remain ambiguous. Recent studies suggest that snow characteristics may be primary drivers of migration, largely due to caribou's high level of mobility and their dependence on landscape conditions for locomotion. We use three datasets to explore the spatial and temporal relationships between landscape-scale caribou movement and snowpack characteristics in the Northwest Territories, Canada, over two decades. We use GPS (Global Positioning System) tracking collar data of barren-ground caribou provided by the Government of the Northwest Territories Department of Environment and Natural Resources to identify individual animal location and migration patterns, with a focus on the Bathurst herd. We derive environmental factors from Calibrated, Enhanced-Resolution Brightness Temperatures (CETB) at 3.125 km resolution and Moderate Resolution Imaging Spectroradiometer (MODIS) data at 500-m resolution. The GPS collar data provides the location of >200 animals (1996-2017) at multiple pre-determined (but varied) time intervals daily. These datasets allow us to further understand whether and how snow characteristics such as melt/refreeze status and the presence of ice are related to caribou movement. Do migration patterns align with snowscape characteristics, possibly indicating a preference for conditions that increase mobility? Do changes in snow factors influence the timing or duration of their movement? A strong relationship between snow melt onset, refreeze cycles, or rain-on-snow events would suggest that expected changes in snowpack variability could lead to range shifts and changes in the overall timing, duration, and synchrony of movement.

[1] Department of Earth and Environmental Sciences, Lehigh University, Bethlehem, PA, USA
[2] Cooperative Institute for Research in Environmental Sciences, University of Colorado Boulder, Boulder, CO, USA
[3] Department of Biological Sciences, Lehigh University, Bethlehem, PA, USA

Local Scale Soil-Snow Interaction and the Impacts of Soil on Snow

MAHSA MORADI[1], JENNIFER JACOBS[1], AND ADAM HUNSAKER[2]

ABSTRACT

The complex interaction between soil and snowpack via energy and mass exchange processes is a critical component of winter at cold regions and has significant consequences for hydrological and ecological cycles. While the impacts of snowpack on the underlying soil and its thermodynamic processes have been widely studied, there are limited studies on soil contributions to the temporal evolution and spatial distribution of snowpack characteristics. The goal of this study is to understand if and how soil exerts controls on snowpack evolution over the course of winter at local scale. The hypothesis is that the soil layers with different physical properties have different thermal and hydrological regimes which impact some characteristics of their overlying snowpack such as its temperature profile. Using observational data at three sites, this study investigates a) if observed depth, temperature, and albedo of snowpacks over different soil layers evolve differently during accumulation and ablation periods and b) how soil physical properties contribute to the temporal evolution of the snowpack characteristics. The study sites, located at the University of New Hampshire Thompson Farm Research Station, Durham, New Hampshire, USA, are selected to represent spatial variability of soil properties and snow depth over the study region. Detailed and continuous observations of meteorological variables, ground heat flux, soil states and snow characteristics including temperature, depth and albedo are collected at these sites to investigate the role of soil layer on snowpack development.

[1] College of Engineering and Physical Sciences, University of New Hampshire, Durham, HN, USA

2 Institute for the Study of Earth, Oceans, and Space, University of New Hampshire, Durham, NH, USA

Remote Sensing Methods for Quantifying Snow Water Equivalent at the Bay of Quite (Ontario) and for Lake Erie Watershed (Ontario)

FÉLIX OUELLET[1] AND AGNES RICHARDS[1]

ABSTRACT

Remote sensing methods used to quantify snow water equivalent in two key priority areas: the Bay of Quinte and the Lake Erie watershed, will be presented. Different remote sensing products and methods have been used to map and identify snow cover. This remote sensing information is cross-referenced with field data and used in running our SNOWPACK model, which uses the following meteorology inputs: precipitation, humidity, wind, and solar radiations. The GIS aspect of the data acquisition and manipulation will also be discussed: spatial references, bandwidth operations, and data format conversions.

[1] Environment and Climate Change Canada, Burlington, ON, Canada

Overview of SnowEx Field Campaign in Northern Alaska, U.S.

CARRIE VUYOVICH[1], SVETLANA STUEFER[2], HANS PETER MARSHALL[3], MICHAEL DURAND[4], KELLY ELDER[5], DRAGOS VAS[6], ARTHUR GELVIN[6], BATUHAN OSMANOGLU[1], CHRISTOPHER LARSEN[7], STINE PEDERSEN[8], DANIEL HODKINSON[1], AND ELIAS DEEB[9]

ABSTRACT

Snow Experiment (SnowEx) was initiated by NASA's Terrestrial Hydrology Program in 2017 to "enable trade studies for a snow satellite mission design." The specific focus of SnowEx is on testing and maturing technology for satellite remote sensing of global snow water equivalent (SWE). Currently, the SnowEx team is planning an airborne and ground-based snow campaign in Alaska in 2022-2023 to address SWE and snow depth measurement questions unique to taiga and tundra snowpacks. Three SnowEx sites are selected in Interior Alaska, a boreal forest environment with discontinuous permafrost and seasonal taiga snowpack. Two SnowEx sites are located on the North Slope of Alaska, a region dominated by low-stature land cover, tundra snowpack, and continuous permafrost. A suite of airborne and ground-based validation activities will take place in fall 2022 and spring 2023 to quantify and compare the capabilities of radar and altimetry sensors to measure differences in SWE and snow depth accumulation during one winter season. The same set of sensors was tested by the SnowEx team in mountain ranges and temperate forests of the Western U.S. in 2017-2021. When taken together, the SnowEx field campaigns provide snow datasets in support of testing and advancement of remote sensing, modeling, and measurements techniques needed for the development of global SWE products. This presentation focuses on the objectives of the boreal forest and tundra SnowEx campaign and presents an overview of upcoming field activities in Alaska.

[1] NASA Goddard Space Flight Center, Greenbelt, MD, USA
[2] Department of Civil, Geological, and Environmental Engineering, University of Alaska Fairbanks, Fairbanks, AK, USA
[3] Department of Geoscience, Boise State University, Boise, ID, USA
[4] School of Earth Sciences, The Ohio State University, Columbus, OH, USA
[5] US Forest Service, Fort Collins, CO, USA
[6] Cold Regions Research and Engineering Laboratory, Fairbanks, AK, USA
[7] Geophysical Institute, University of Alaska Fairbanks, Fairbanks, AK, USA
[8] Cooperative Institute for Research in the Atmosphere, Colorado State University, Fort Collins, CO, USA
[9] US Army Engineer Research and Development Center, Cold Regions Research and Engineering Laboratory, Hanover, NH, USA

Using Unpiloted Aerial Systems Structure from Motion Photogrammetry for Avalanches in Mount Washington, NH

CAMERON WAGNER[1], ADAM HUNSAKER[1], TIM HOHENEDER[2], AND JENNIFER JACOBS[3]

ABSTRACT

Unpiloted Aerial Systems (UAS) collected structure from motion (SfM) photogrammetry is a proven effective high resolution snow depth measurement technique with a potential for avalanche forecasting. In winter 2021/22, a series of flights were conducted in Tuckerman Ravine and Boott Spur Gullies avalanche zones on Mount Washington, NH, USA. The goal of these flights was to: (1) quantify wind slab depth across varying aspect terrain, (2) model wind slab distribution over the same terrain given actual meteorological conditions, and (3) engage with the Mount Washington Avalanche forecasters and backcountry skiers. Mount Washington east aspect glacial cirques were chosen for this project due to their predominant wind slab avalanche problem attributed to common hurricane force winds and ample snowfall in fetch areas above the cirques. Ground truthing was conducted to quantify the SfM effectiveness in measuring snow depth in this hostile mountain environment. It is found that UAS collected SfM photogrammetry is effective in quantifying wind slab depth with a 5 cm spatial resolution. Qualitative methods and results from stakeholder engagement will also be presented.

[1] Institute for the Study of Earth, Oceans, and Space, University of New Hampshire, Durham, NH, USA

[2] Natural Resources and Earth System Science, University of New Hampshire, Durham, NH, USA

[3] College of Engineering and Physical Sciences, University of New Hampshire, Durham, HN, USA

Tundra Snow Depth Retrievals from Wideband Radar Observations in Trail Valley Creek

WEI WANG[1], RICHARD KELLY[1], AND JOSHUA KING[2]

ABSTRACT

Recent experiments have explored the potential of using airborne wide-band snow radar observations to estimate snow accumulation. Most if not all experiments have focused on sea ice applications with a series of campaigns conducted by Operation Ice Bridge (OIB) successfully demonstrating the applicability of wideband observations (2-8 GHz) to estimate snow on sea ice. Most applications of the wideband approach for terrestrial snow applications have been conducted using ground-based systems. However, in 2019 a wideband radar system (2-18 GHz) was flown over Trail Valley Creek in the North West Territories, Canada coinciding with ground-based field measurements of snow depth and SWE properties from a range of measurement systems. This paper describes the application of the using the interface-based pulse peakiness snow depth retrieval method to estimate snow on land. The approach was tested on the six major vegetation types present at the study area (white and black spruce trees, tall shrub, riparian shrub, dwarf shrub, tussock, and lichen). Snow depth derived from Airborne Laser Scanner (ALS) point clouds was used as the reference for snow depth retrieval. To address the differences between snow on land and on sea ice, a recalibration of the algorithm parameters was completed as well as introducing additional measures to screen abnormal observations with a low Signal to Noise Ratio. It was concluded that the principles behind the pulse peakiness approach was generally valid for tundra snow depth estimates at study sites characterized by smooth and low slope surfaces and with low stand vegetation. The presence of surface vegetation and increased surface roughness led to increased complexity of radar waveforms, which resulted in increased snow depth retrieval uncertainty.

INTRODUCTION

Seasonal snow is an essential component of the hydrologic cycle, where changes in snow cover and distribution can have significant impacts on precipitation, wildlife, and water supply. In addition, variations in snow cover can serve as an indicator for climate change (Croce *et al.*, 2018). A timely accurate estimate of snow depth or snow water equivalent (SWE) can be used for hydrology and climate modeling, managing local water supply, as well as presenting a quantifiable parameter that responds to climate change.

It is because of the necessity of snow monitoring that the methods for snow depth / SWE estimates have been developed throughout the years. While field measurements or records from weather stations are known to be reliable, they are localized in representativeness and so to cover the vast areas around the globe, remote sensing technology including passive and active microwave sensors have been utilized along with numerical models.

---

[1] Department of Geography and Environmental Management, University of Waterloo, Waterloo, ON, Canada
[2] Environment and Climate Change Canada, Toronto, ON, Canada

More specifically for snow depth or SWE retrieval using active sensors, there have been experimental observation methods developed utilizing radar systems at Ku or X band (Yueh *et al.*, 2009; Oveisgharan *et al.*, 2020; Leinss *et al.*, 2015). As explained by Leinss *et al.* (2015) and Oveisgharan *et al.* (2020), it is possible to retrieve SWE with the interferometric phase differences between two measurements at specific frequencies, especially at L-band. Cui *et al.* (2016) proposed using Synthetic Aperture Radar (SAR) observations at Ku and X band coupled with a backscatter forward model to estimate the optical thickness of the snowpack. By contrast, Operation Icebridge (OIB) deployed an experimental wide band radar with a frequency range of 2-8 GHz which differs from the radar systems used by Leinss *et al.* (2015) and Cui *et al.* (2016) in that the frequency of the transmitted signal is constantly modulated. In addition, various other FMCW radar systems have also been used for SWE retrieval and snow stratigraphy measurements (Marshall and Koh, 2008). However, as those mostly ground-based FMCW radar systems are not the focus of this paper, a wide band radar here is specifically defined as an airborne FMCW radar capable of operating in a wide bandwidth similar to the ones used in OIB and AWI Icebird campaigns.

Currently, most snow depth retrieval algorithms using data collected by a wide band radar are based on obtaining two crucial interfaces, air-snow interface, and snow-ice / snow-ground interface (Kwok *et al.*, 2017). As OIB which provided a large percentage of the data used for developing various retrieval algorithms is focused on offering detailed information about polar ice, most of these retrieval methods are for snow on sea ice. According to Kwok *et al.* (2017), the difference amongst these algorithms mainly lies in how the two interfaces are obtained and how they are optimized for a specific radar system. Kurtz and Farrell (2011) and Kurtz *et al.* (2013) proposed an interface identifying method that combined searching for local maximum and empirical thresholding, which was adopted by the National Snow and Ice Data Center (NSIDC). Later on, a waveform fitting method was developed by Kurtz *et al.* (2014) where a physical model was established to simulate wide band radar waveforms over sea ice and the two interfaces were obtained by fitting data to the model. In addition, a retrieval based on the gradient of radar intensity return was adapted from Koenig *et al.* (2016) and tested in Kwok *et al.* (2017). In comparison, Newman *et al.* (2014) took a slightly different approach that avoided the need for fixed sets of thresholds used by several previously mentioned methods. As demonstrated by Newman *et al.* (2014) and Kwok *et al.* (2017), a Haar-CWT was utilized to detect abrupt changes within the waveform and those identified abrupt transitions were used to determine the two interfaces under specific criteria. To cope with the gradual improvement of the radar system used in different campaigns in OIB, a simplified version of the algorithm from Kwok and Maksym (2014) was developed where interfaces were located by measuring their deviation from their neighbors. Kwok *et al.* (2017) concluded that all methods described above were capable of reproducing the high snow depth pattern expected in the study area, but the performance of each individual method when compared with field measurement differs. More specifically, according to Kwok *et al.* (2017), when comparing the retrieval snow depth from these five methods with field data from Eureka, their corresponding mean bias are -0.8 cm (Kurtz *et al.*, 2013), -5.7 cm (Kurtz *et al.*, 2014), 1.3 cm (Koenig *et al.*, 2016), 2.0 cm (Newman *et al.*, 2004), and -2.2 cm (Kwok and Maksym, 2014)

Since the start of OIB, most algorithms developed including the ones described above are for snow on sea ice because of several possible reasons. First, the difference between dielectric properties of snow and frozen soil is generally weaker compared with snow and sea ice; second, the presence of vegetation presents a sizable challenge. When there is vegetation buried in the snowpack, it would likely increase the complexity of the snowpack while if the vegetation is emerging from the snowpack, it could interfere with radar waveforms before they reach the snowpack. Thirdly, terrain features on land like steep slopes and hummocks could further increase the complexity of the waveforms. In addition, the frequency ranges of the radar systems used by these algorithms are often not wide enough to cover shallow snow. However, Icebird 2019 Winter collected data over the Trail Valley Creek research station using a wide band radar with a frequency range of 2-18 GHz and this provides an opportunity to examine waveforms for snow over various land terrain features including different types of vegetation as well as hummocks and to determine whether principles behind snow on sea ice algorithms can be used for snow on land.

STUDY AREA

The study area for this experiment is based at the Trail Valley Creek (TVC) research station, located at the northern boundary of the tundra-boreal forest ecotone, roughly 50 km north of Inuvik, Northwest Territories (Pohl *et al.*, 2005; Pomeroy *et al.*, 1998). The TVC research basin is within the Inuvialuit Settlement Region near the Inuvik-Tuktoyaktuk Highway. The watershed drains approximately 58 square kilometers into the Husky Lakes Estuary network. In addition, the basin itself is underlain with a continuous ice-rich permafrost.

Topographically, the area is dominated by gently rolling hills (Pohl *et al.*, 2005). A major terrain feature of the study area is the hummocks, which are defined as sub-meter to meter-scale circular to oval mounds (Verret *et al.*, 2019). The existence of hummocks leads to a relatively large amount of the micro terrain that is below the spatial resolution of most field measurements. More complex terrain is present in the southern margins of the study area, in proximity to forests.

According to Boike and Grünberg (2019) and Palmer *et al.* (2012), the study area contains 6 main vegetation types including trees (white and black spruce (Picea glauca and Picea mariana), tall shrubs, riparian shrubs, dwarf shrubs, tussocks, and lichens. Tussocks, lichens and dwarf shrubs are the dominant vegetation types while trees and tall shrubs have a relatively sparse distribution. For tall shrubs particularly, there are often other low height vegetation in between two tall shrubs (Boike and Grünberg, 2019). The landscape also features several lakes as shown in Figure 1 with a number of streams and rivers running through the study area.

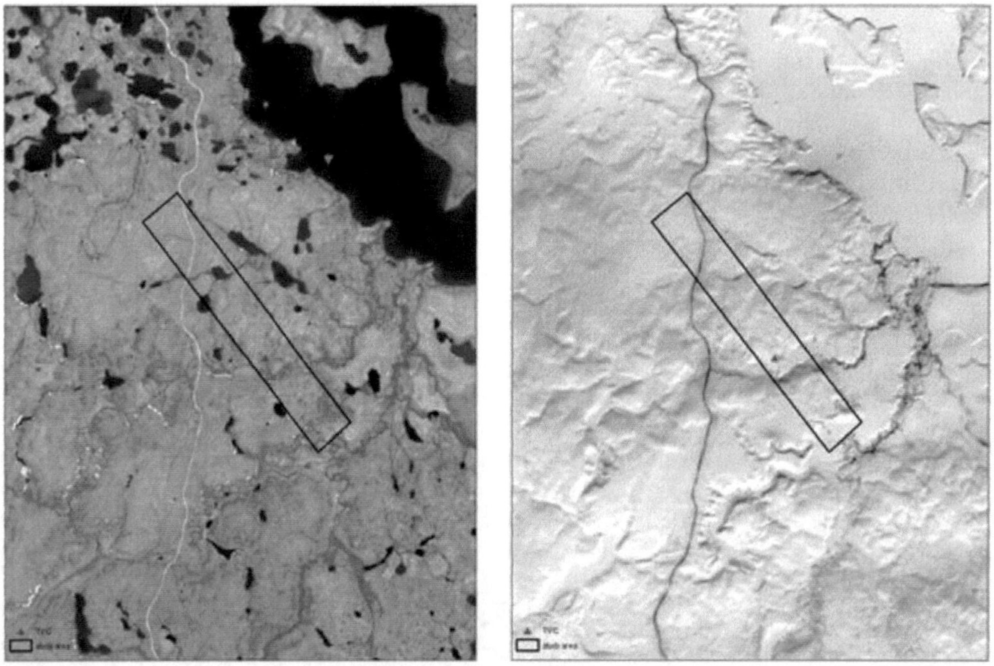

Figure 1. Landsat 8 images of study area; left is from 2019.06.18 and right is from 2019.04.15. Red triangle represents the location of TVC research station.

DATASETS

Snow Radar Data and Airborne Laser Scanning (ALS) Data from Icebird 2019 Winter

The Icebird 2019 Winter campaign was aimed at conducting sea ice surveys over different ice regimes to observe, analyze, and estimate sea ice thickness in addition to examining regional snow distribution (Hendricks *et al.*, 2019). An airborne wide band snow radar system observed snow on sea ice and an ALS system observed surface roughness features (Hendricks *et al.*, 2019). Five field observation periods were planned for the 2019 winter campaign, in which the fourth period, conducted between April 6th and April 11th, 2019, was when the snow radar and ALS data were collected and are used in this study. The particular study date that covered the TVC region was April 10th, 2019.

The onboard ALS system was a Riegl VQ-580 (Hendricks *et al.*, 2019). The ALS point cloud data used in this study was level 1b with the included metadata consisting of geolocation and positioning information. With AWI toolbox (Alfred Wegener Institute, Helmholtz Centre for Polar and Marine Research, 2019), it was possible to filter the large point clouds by time period and narrowed it down to point clouds that were within the TVC study area.

The snow radar system, mounted on the Polar-6 Basler BT-67, aircraft was set to operate in Frequency-Modulated and Continuous Wave (FMCW) mode for low-altitude missions. The aircraft flew at an elevation of approximately 61 m (200 ft) and 488 m (1600 ft) when using the radar for observation (Hendricks *et al.*, 2019). The wide band radar operated at a frequency between 2 and 18 GHz with 2 receiving channels (horizontal and vertical). Four combinations of polarization observations were achieved, VV, VH, HV, HH. The theoretical range resolution within the snowpack, assuming a snow density of 300 kg m^3, was 1.14 cm. At 61 m (200 ft) height, the radar had an across-track footprint of 2.1 m and an along-track footprint of 2.0 m while at 488 m (1600 ft) height, the two respective footprints were 6.0 m and 10.8 m. According to Hendricks *et al.* (2019), initial processing of the raw wide band radar was done using CRESIS Matlab Toolbox, followed by coherent noise removal and deconvolution of echograms (Julita *et al.*, 2021).

After the pre-processing, the data was separated in to 24 continuous segments of the flight path in a Matlab file format. Within the TVC study area, elements from 6 different flight paths were used (path numbers 1, 2, 6, 7, 11, 12, at VV polarization). The reason for using VV polarization instead of HH is that VV waveforms are generally of higher quality, for example, less noise and interference, and are more suited for identifying peaks. Using PySnowRadar (King *et al.*, 2020), the files were read into python and each waveform was presented as a 1x14050 numpy array, with each element of the array being an intensity return. Each waveform data record included the impulse data, longitude, latitude, and the elevation of the aircraft. Figure 2 shows the flight line locations. Waveforms over water bodies (lakes and rivers or streams) were removed from the analysis leaving a total of 3128 radar data locations.

ALS Point Clouds From 2016

A snow-free ALS data set for the study region was acquired on September 13th, 2016, using a Riegl LMS-Q680i. The reason for needing this flight was to provide a snow-off DEM for the development of a lidar snow depth product when combined with a snow-on DEM from 2019. Details of data acquisition can be found in Anders *et al.* (2016). Three products from Anders *et al.* (2016) were used in this study including a digital terrain model (DTM), maximum vegetation height map and mean vegetation height map, as is shown in Figure 3. All three data have a spatial resolution of 1 m and were later clipped to the study area. The 2016 DTM was used as a snow-off reference for the 2019 snow-on DTM to generate a snow depth map. Both vegetation height maps were used in conjunction with the TVC vegetation map to provide a guideline for typical heights of different vegetation types.

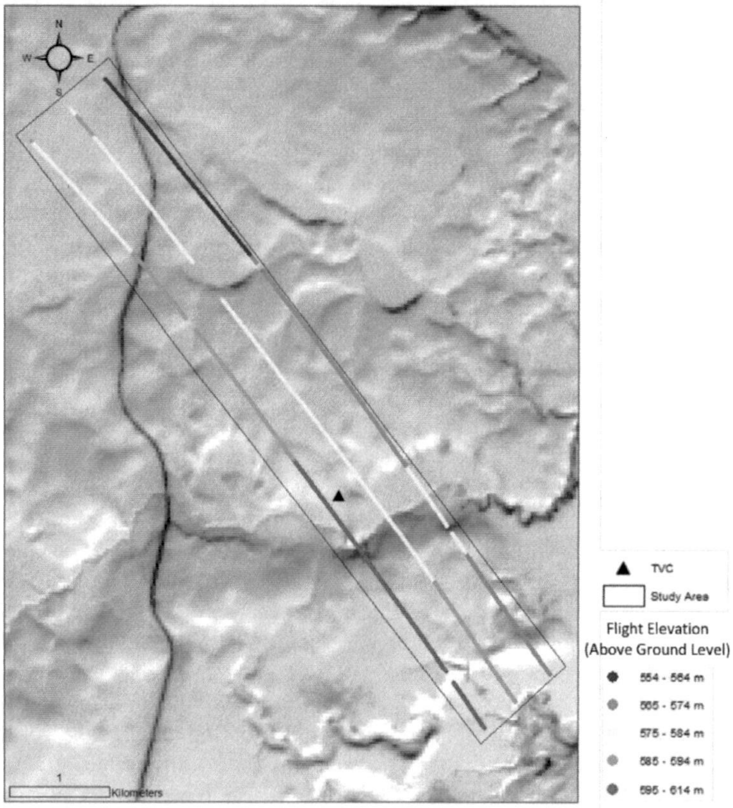

Figure 2. Location of the used radar data points colored by flight elevation (Landsat 8 from 2019.04.15 as background).

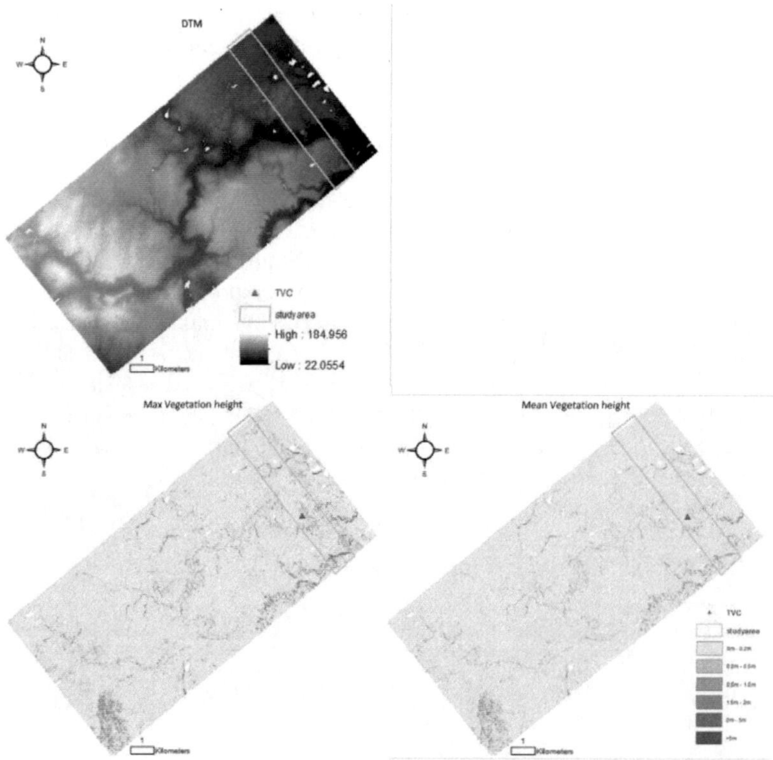

Figure 3. 2016 snow off DTM, max vegetation height and mean vegetation height maps.

Field Snow Depth Probe Measurements

Field snow depth data were collected during three field campaigns in the winter of 2018-2019. Snow measurements focused on the open tundra area around the TVC research station. Snow depth measurements were made using a GPS-enable depth measuring probe, called a Magnaprobe (SnowHydro, n.d.). Table 1 shows the dates of the three campaigns as well as the number of depths recorded. In this paper, only the data from the last campaign were used because this campaign was temporally closest to when the radar flights took place on April 10th. Due to the limited coverage of the 2019 ALS point clouds, only approximately half of the measurements from the third Campaign were used (4334 points), whose locations and snow depth distribution are shown in Figure 4 and Figure 5. Most points have a snow depth between 0.3 and 0.5 m. The mean snow depth for all the used measurements is 0.42 m with a standard deviation of 0.13 m. As the field measurements do not cover gulley where deep snow is most likely to accumulate, there are few points with snow depth higher than 0.9 m.

Table 1. Details of three field campaigns.

	Dates	Number of Depths Recorded
First Campaign	November 12th -18th 2018	6185
Second Campaign	January 11th-20th 2019	6740
Third Campaign	March 19th-27th 2019	8541

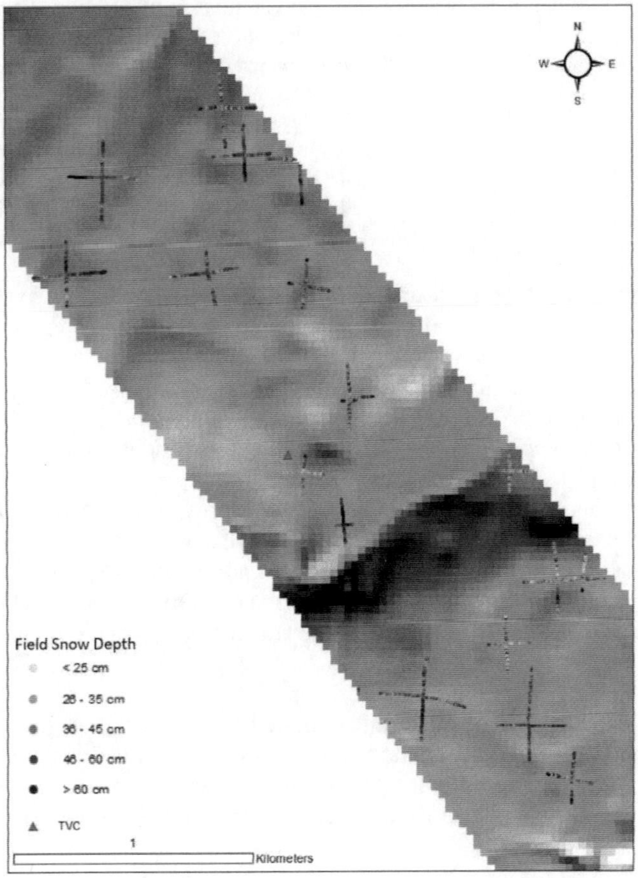

Figure 4. Locations of used field measurement (Landsat 8 from 2019.04.15 as background).

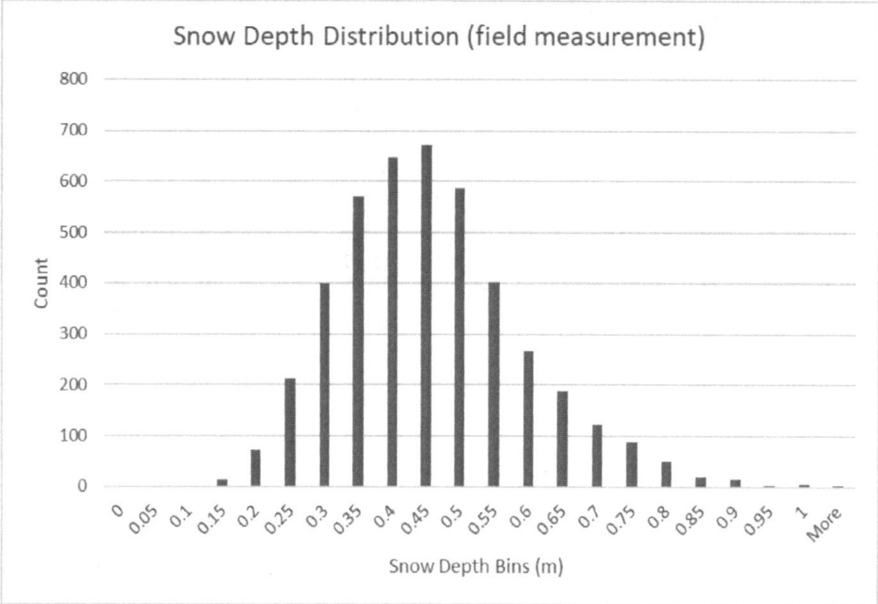

Figure 5. Histogram distribution of snow depth measurements (collected between 2019.03.21 and 2019.03.26).

In addition, Table 2 shows the mean and median snow depth and its standard deviation for each vegetation type from the field data. Among the six vegetation types, tree and tall shrub have the deeper snow but also a larger standard deviation while dwarf shrub, tussock and lichen are comparable in terms of mean, median snow depth and standard deviation.

Table 2. Mean and median snow depth and standard deviation for each vegetation type.

	Tree	Tall Shrub	Riparian Shrub	Dwarf Shrub	Tussock	Lichen
Mean (m)	0.56	0.63	0.49	0.42	0.42	0.43
Median (m)	0.56	0.65	0.47	0.41	0.41	0.43
Std (m)	0.21	0.16	0.17	0.13	0.13	0.12

TVC Vegetation Map

To account for the effect of different underlying vegetation on the radar waveform, a vegetation map was used to analyze spatial variability of vegetation type in the study region. A vegetation map of TVC was provided by Boike and Grünberg (2019), with a spatial resolution of 10 m. Vegetation is classified into 6 classes including tree, tall shrub, riparian shrub, dwarf shrub, tussock, as well as lichen as is shown in Figure 6. The map also contains a seventh class, water. The map covers the same area as the 2016 ALS point clouds data introduced above and is clipped to the study area.

According to Boike and Grünberg (2019), the vegetation map was generated through combining ALS point clouds and airborne orthophotos. Reference areas were chosen and vegetation types in these areas were classified manually, after which a set of parameters were calculated to characterize each vegetation type. Parameters include terrain attributes derived from ALS point clouds (e.g., slope, aspect, vegetation height, *etc.*) as well as spectral attributes from orthophotos (e.g., average hue, average saturation, *etc.*).

In their study, Boike and Grünberg (2019) reported a final map accuracy of 87.7% for the validation dataset and 89.5% for the calibration dataset. However, it was noted that with a 10 m spatial resolution grid, mixing of different vegetation types was likely in a single cell, specifically for tall shrub, where the actual tall shrubs are sparse and between the tall shrubs are normally low vegetation.

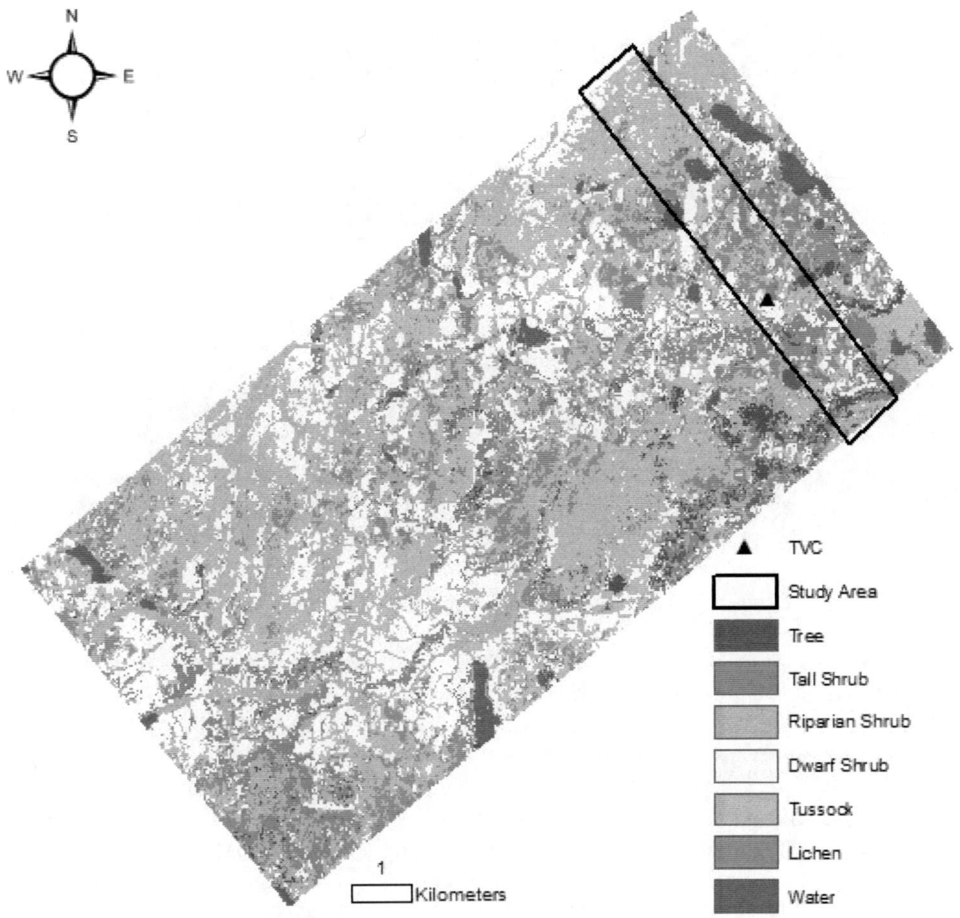

Figure 6. TVC vegetation map (Boike and Grünberg, 2019).

METHODS

Pre-processing

Data pre-processing was divided into 2 components: the generation of a snow depth product for April 2019 and radar waveform pre-processing. Because the sparse field snow depth measurements did not spatially match the flight path of the airborne wide band radar, to provide snow depth reference for each radar data location, a snow depth product was generated by differencing the snow-off DTM from September 13th, 2016 from the snow-on DTM from April 10th 2019. In addition, based on the snow depth distribution from the field data, any snow depth that is above 2.5 m or a negative value is treated as invalid (NaN).

Four pre-processing steps were applied to radar waveform to prepare them for analysis. First, for each radar waveform (14050 range bins), the first 500 and the last 500 bins were removed to eliminate residual noise from these range bins, as is shown in Figure 7. More specifically, the maximum return of each waveform is assumed to be the snow-ground interface because the difference of dielectric properties between snow and soil is theoretically the largest amongst all different interfaces. However, for some waveforms, the maximum return is located in the beginning range bins or in the final range bins and is an artefact of system processing rather than geophysical signal since both sets of range bins are located geometrically far from where the snow-ground interface is expected.

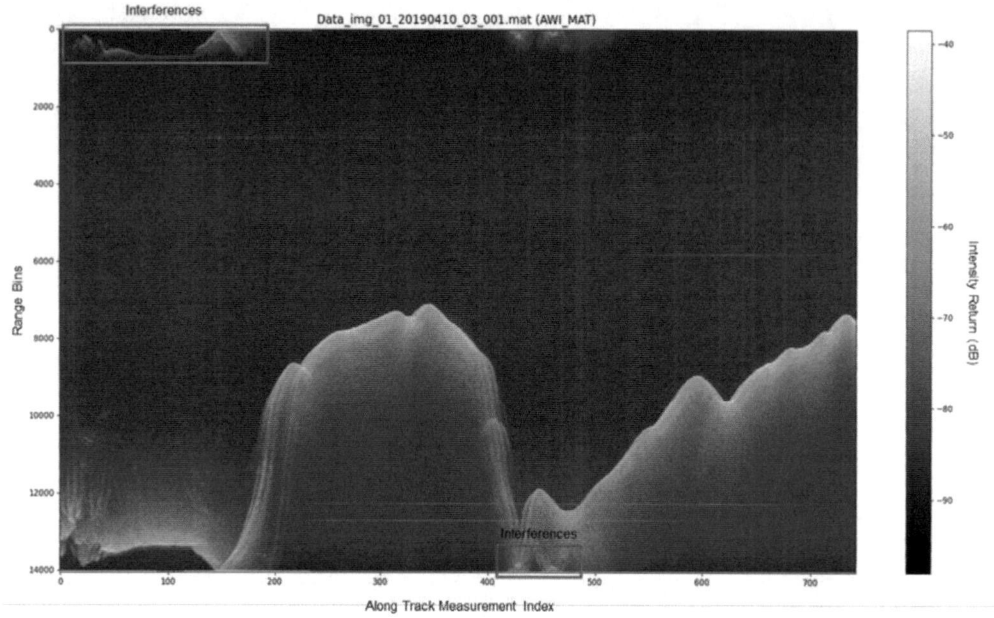

Figure 7. Interference in the beginning and final bins.

Second, a mean noise value was recorded for each waveform by averaging the first 100 returns after the first step. This was used as base reference to distinguish an interface peak in the waveform from noise. Third, the waveform was further reduced in length to 300 range bins, 150 bins before and 149 bins after the waveform maximum. The waveform maximum response is assumed to be the ground surface (snow-ground interface) which has the strongest power return. The reason for centering each waveform on its maximum was to provide a consistent reference when comparing different waveforms, regardless of ground elevation and flight elevation. However, the number of bins is arbitrary, and is largely influenced by the distribution of snow depth. Fourth, the waveforms were linearly normalized by dividing each return by its maximum response.

Evaluation of Lidar Snow Depth Product

Field snow depth data from King *et al.* (2020) collected between 2019.03.21 and 2019.03.26 were used as a ground reference to evaluate the lidar snow depth product obtained from differencing the 2016 and 2019 DTMs. The 1-m cells of the lidar product were converted into a point shapefile using the centroid of each cell. For each field snow depth measurement, its coincident lidar snow depth was obtained by selecting the closest converted lidar snow depth centroid within 1 m as is shown in Figure 8. Comparisons of field and lidar snow depths were conducted for each vegetation type. Distribution parameters for both data were calculated (mean, standard deviation, and median) in addition to obtaining mean absolute error (MAE).

Estimation of Snow Depth Using the Pulse Peakiness Approach

Julita *et al.* (2021) described the pulse peakiness approach as one of a range of peak picking methods based on the pulse peakiness of local peaks. Most of these peak picking methods have been applied for sea ice applications (Kwok *et al.*, 2017; Newman *et al.*, 2014; Kurtz *et al.*, 2013). As a peak in the waveform can potentially indicate the boundary between two mediums with different dielectric properties, the approach was essentially searching for peaks that represent the air-snow and the snow-ice interfaces. Logarithmic scale of the impulse return was used for locating the air-snow interface and linear scale of the intensity returns was used for obtaining the snow-ice interface. The reason for using linear for the snow-ice interface is that the difference of dielectric properties between snow and sea ice is large enough that using a linear scale will decrease the number of peaks detected while preserving the interface peak.

Figure 8. Matching field snow depth and lidar snow depth.

Eligible peaks were selected by applying a threshold over all detected peaks based on manually input parameters (log threshold & linear threshold) as well as the mean noise. A screening process was applied after identifying all the eligible peaks: if in linear scale, there were more than 5 eligible peaks, the waveform was considered too complicated to produce a robust result. For waveforms with fewer than 5 linear eligible peaks, two pulse peakiness values were calculated for each eligible peak. Unlike the pulse peakiness used in sea ice classification, Julita *et al.* (2021) defined these two pulse peakiness (left and right) as:

$$PP_l = \frac{S_{peak}}{\frac{1}{N}\sum_{i=1}^{N} S_{peak-i}} \times N$$

$$PP_r = \frac{S_{peak}}{\frac{1}{N}\sum_{i=1}^{N} S_{peak+i}} \times N$$

where S_{peak} is the linearly normalized intensity return, N is the bin range before and after the local peaks indicating how many bins are included into the calculation. It was recommended to use $N = 10$ because it was 2 times the range resolution of the radar (Julita *et al.*, 2021). Finally, air-snow interface was defined to be the first eligible peak whose PP_l exceeded a set threshold while snow-ice interface was defined to be the last eligible peak whose PP_r exceeded another set threshold.

Two changes were made to the pulse peakiness method that was developed for snow on sea ice. First, the process of defining the snow-ground interface using pulse peakiness "right" was removed and instead, it was assumed that the strongest return would be the snow-ground interface. For sea ice applications, there may be circumstances where the strongest return was the air-snow interface due to high brine concentrations in the lower part of the snow. Geldsetzer *et al.* (2019) has shown that the C-band radar response from snow on first year sea ice is complicated by the brine layer. The effect of the brine layer on other frequencies as wide band radar covers from 2GHz to 18GHz is less well described and possible future work is needed. Pulse peakiness right is introduced to address the situations where air-snow interface is the strongest return. However, for terrestrial snow, this is irrelevant. Second, additional screening process were added to overcome the complexity of the waveforms. Compared with snow on sea ice, it is expected that there are considerably more complex reflections in waveforms on land especially due to complicated vegetation interactions. Two kinds of complex interaction are expected: interactions between the wave, snow, and emergent vegetation such as trees and tall shrubs, and interactions between snow and buried vegetation such as dwarf shrubs. Therefore, in addition to the original screening where the waveform must have no more than 5 eligible linear peaks to have a valid output, 3 more restrictions were added. To begin with, the normalized linear return of the air-snow interface peak must be less than 0.6. Moreover, the air-snow interface must be ahead of the strongest return (assumed snow-ground interface) in range bin sequence. Finally, a target peak was introduced to further filter out the waveforms that were too sophisticated. A target peak was defined to be the closest eligible logarithmic peak to the reference air-snow interface, where the reference air-snow interface was obtained by converting the lidar snow depth into range bins using an assumed snow density (Fig. 9). For any waveform, if no target peaks were present within 20 range bins of the reference air-snow interface, then the waveform was considered to be ambiguous and was removed from the analysis.

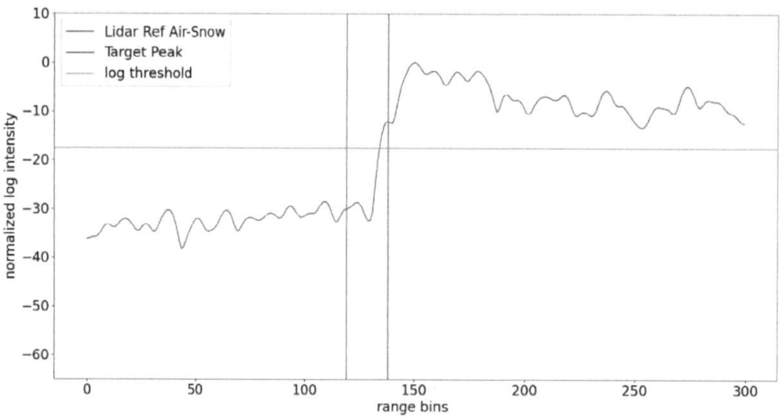

Figure 9. Example target peak defined by the closet eligible log peak to the lidar reference air-snow interface (waveform shown is from tussock).

In summary, the snow-ground interface was assumed to be the strongest return in the waveform. Therefore, the algorithm required 3 parameters instead of the 4 originally defined by Julita *et al.* (2021): 1) logarithmic threshold for determining eligible peaks in logarithmic scale; 2) linear threshold for determining eligible peaks in linear scale, and 3) the left pulse peakiness threshold for choosing the air-snow interface. All three parameters were optimized by using lidar snow depth data as reference. For both logarithmic and linear threshold, the range was set to be between 0.1 and 0.9; for the left pulse peakiness threshold, the range was from 10 to 35. Optimization scoring for each set of parameters is shown below:

$$score = 0.75 \times MAE - 0.25 \times number\ of\ valid\ outputs$$

Both MAE and the number of valid outputs were normalized to ensure they were similarly scaled (from 0 to 1). In addition, to ensure there would be sufficient points for evaluation, a requirement for the number of valid outputs under any set of parameters must be above 10% of the total points was added. Optimization was performed for each vegetation type separately. Finally, the best set of parameters were defined as the ones with the lowest score.

The optimization algorithm was run using 3 different divisions of the dataset. First, it was run on the entire dataset, and then the optimal parameters were applied to the entire dataset, which represented a best-case scenario. Second, it was run on an 80-20 split of the dataset; 80% of the data were used for optimization/training and 20% of the dataset for testing. Third, a 50-50 split of the dataset, 50% optimization/training and 50% testing.

RESULTS

Evaluation of Lidar Snow Depth Product

Figure 10 shows the 1-m spatial resolution lidar snow depth product from the 2016 snow-off DEM and 2019 snow-on DEM. The snow depths were around 30-40 cm in relatively flatter grounds mostly covered by tussocks, lichen, and dwarf shrubs. This is similar to the distribution of field snow depth measurements despite there being a two-week gap between when the field measurements and wideband radar measurements were made. In areas near streams or some valleys in the bottom part of the study area, the snow depth was greater than 60 cm, as indicated by the bright white in Figure 10. This is due to the collection of snow in these gullies and depressions.

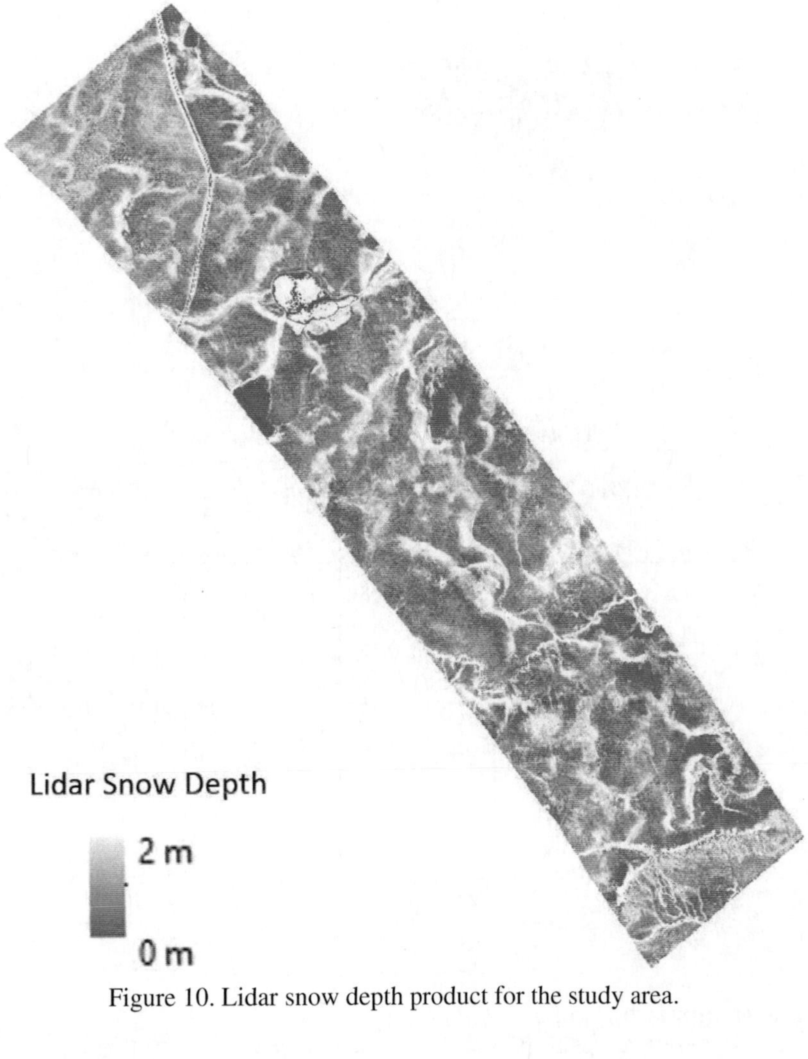

Figure 10. Lidar snow depth product for the study area.

There were 4334 field snow depth measurements used for the evaluation of lidar snow depth product. Table 3 shows the results of comparing the lidar snow depth with field measurement. Treating field snow depth as the reference, the lidar estimates underestimate snow depth for all six vegetation types, especially for tree, riparian shrub, and dwarf shrub classes as is shown in Figure 11. It is necessary to note that there was a 14-day difference between the time when these two data were collected. According to local weather station data, the temperature remained below zero with no significant precipitation event, however, some snowmelt might have been present in the field which could explain part of the reason for the underestimation. Further work to verify this possibility is ongoing. In addition, it is possible that the underestimation was the product of mismatch between snow-off and snow-on dates (the snow-off DEM used was from 2016); the underlying terrain or vegetation may have changed, leading to an inaccurate lidar snow depth. Equally, there could be an over-probe of up to 7.6 cm for the magnaprobe, as is mentioned in King *et al.* (2020), which would contribute to field snow depth being relatively greater.

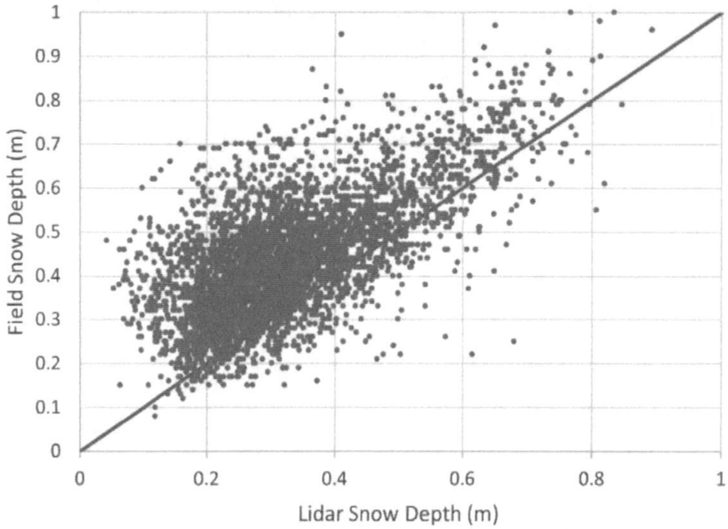

Figure 11. Scatterplot comparing lidar and field snow depth.

Table 3. Evaluation of lidar snow depth.

	Tree		Tall Shrub		Riparian Shrub		Dwarf Shrub		Tussock		Lichen	
	Field	Lidar	Field	Lidar	Field	Lidar	Field	Lidar	Field	Lidar	Field	Lidar
Mean (m)	0.56	0.45	0.63	0.53	0.49	0.38	0.42	0.28	0.42	0.33	0.43	0.35
Std (m)	0.21	0.20	0.16	0.17	0.17	0.16	0.13	0.11	0.13	0.11	0.12	0.12
Median (m)	0.56	0.39	0.65	0.57	0.47	0.34	0.41	0.26	0.41	0.31	0.43	0.33
RMSE (m)	0.19		0.15		0.17		0.19		0.13		0.13	
Bias (m)	-0.11		-0.10		-0.11		-0.15		-0.09		-0.08	
Std_bias (m)	0.16		0.11		0.13		0.12		0.09		0.10	
Count	13		151		231		1316		1456		1167	

Pulse Peakiness Approach Results

Table 4 shows the optimal pulse peakiness parameters for each vegetation type using a 100-0 split, and Figure 12 shows the retrieved snow depth variation for each vegetation classes. The pulse peakiness estimates used are from the optimal parameters obtained for a 100-0 split. The reason for using a 100-0 split is that the randomness in splitting the data has a significant effect on the optimization results, especially for tree, tall shrubs, and riparian shrubs. Moreover, a 50-50 split may result in insufficient amount of the waveforms for the algorithms, which will enhance the effect of the randomness in splitting. It is necessary to note that this is the first attempt of developing the method and a stratified random split could be tested in the future.

Table 4. Optimal pulse peakiness parameters for each vegetation type

	Log	Lin	PP_L	MAE (m)	Bias (m)	RMSE (m)	Pearson	Valid/Total Points
Tree	0.6	0.5	25	0.12	0.03	0.15	0.65	11/84
Tall Shrub	0.3	0.8	20	0.27	0.13	0.34	0.20*	226/441
Riparian Shrub	0.2	0.7	15	0.29	0.21	0.38	0.35*	209/321
Dwarf Shrub	0.4	0.5	25	0.13	0.04	0.19	0.29*	135/500
Tussock	0.5	0.5	30	0.09	0.04	0.12	0.55*	141/1224
Lichen	0.3	0.5	30	0.10	0.01	0.15	0.46*	74/558

*indicates significant P-value at α<0.01

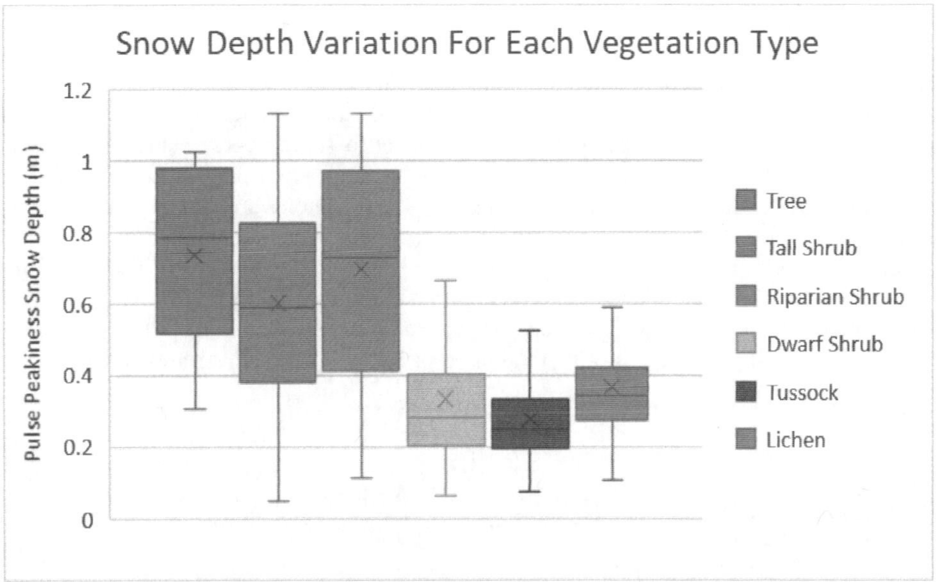

Figure 12: Pulse peakiness snow depth variation for each vegetation type

Overall, from Figure 12, snow depth variation is considerably higher for trees, tall shrubs and riparian shrubs in addition to them having generally thicker snow, possibly due to the higher vegetation.

For trees, despite a relatively high Pearson correlation coefficient and a low MAE, the *p*-value is not significant because there are few points. The scatterplot (Figure 13a) for trees further proves the high correlation and low MAE where most points fall closely around the 1:1 line. While this is promising, due to the sparse distribution of trees in the study area, the valid points shown here are likely to be returns from spaces between trees instead of actual tree returns. This is further reinforced by the similarity of parameters between tree and three other low height vegetation (dwarf shrub, tussock, & lichen). The low linear threshold at 0.5 filtered out all waveforms from actual tree returns since these waveforms are highly complex with a large number of peaks. Both tall shrub and riparian shrub have low Pearson correlation coefficients (and high MAE values) of 0.2 (0.27m) and 0.35 (0.29m), respectively. The scatterplots (Figure 13b and c) show the characteristics of relationships between radar and reference estimates with no discernible relationship between tall shrub and lidar and a general overestimation of the radar estimates for tall shrubs. An additional observation here is the high linear threshold of these two vegetation types compared with low height vegetation like tussock. As a high linear threshold leads to complex waveforms (ones often challenging for this retrieval approach) being more likely to be fed into the retrieval, the optimal linear threshold should be as low as possible provided that there will be sufficient valid output. For tall shrub and riparian shrub, a high optimal linear threshold indicates that fewer waveforms from these two vegetation types resembles the ideal two-peak shape, and the optimization was forced to include more complex waveforms to satisfy the criteria of having enough valid outputs.

For dwarf shrub, the MAE is 0.13 m with a Pearson correlation coefficient of 0.29. Due to the lower linear threshold, the increase in total points does not lead to an increase in valid points. From Figure 13d, the reason for a low Pearson correlation coefficient is the overestimation of snow depth by the pulse peakiness algorithm. More specifically, there are a series of points with increasing pulse peakiness snow depth, whose Lidar snow depths remain constant at around 0.2 and 0.3 m. This is most likely to be caused by the waveform having multiple peaks with close to maximum returns and will be further discussed later. In comparison, the underestimation of the retrieval retained the observable pattern from riparian shrub as most of the points above the 1:1 line still falls into a linear pattern.

According to Boike and Grünberg (2019) tussock and lichen classes are characteristically similar in classification terms. However, they are functionally different biophysically. They behave similarly in terms of their radar waveform. Compared with other vegetation types, they have the best retrieval results with MAE of 0.09 m for tussock and 0.1 m for lichen and Pearson correlation coefficients of 0.55 and 0.46, respectively. This result is to some degree achieved by severely constraining the waveform selection for use with the algorithm, as indicated by the low valid / total points ratio. In other words, for tussock and lichen, the two thresholds are relatively more effective at filtering out waveforms that are difficult for the algorithm to accurately obtain interfaces. Both the decrease in MAE and the increase in Pearson correlation coefficient are illustrated in the scatterplot of Figure 13e and f, where the overestimation that results in the low correlation for dwarf shrubs has mostly disappeared. It is noted that the pattern for underestimation of the method persists in both of these vegetation type, indicating that this could possibly be addressed with empirical bias correction.

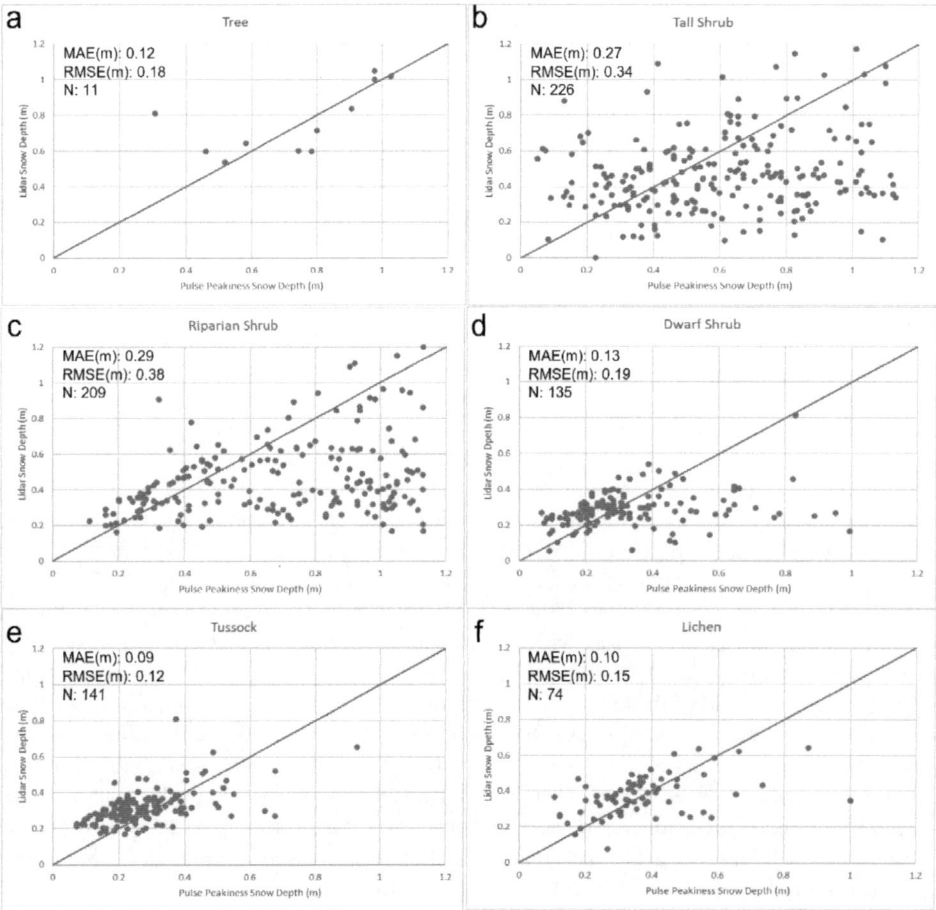

Figure 13. Scatterplots between pulse peakiness and lidar snow depth estimates for the following classes a) tree, b) tall shrub, c) riparian shrub, d) dwarf shrub, e) tussock and f) lichen.

DISCUSSION AND CONCLUSION

Underestimation and Overestimation

Both underestimation and overestimation exist but they result from different sources. With respect to underestimation, it is likely to not be from waveform / radar measurement and it follows a clear pattern, meaning that it can be mitigated to a degree with a bias correction. On the contrary, the overestimation is mainly resulted from the waveforms and needs other more delicate solutions.

The tussock land cover class is used to illustrate the underestimation problem as this landcover class has the most radar points. When running the pulse peakiness approach on the 100-0 split scenario, out of 141 valid outputs, 57 had an absolute error of 10 cm or above. And out of these 57, 39 underestimated snow depth in a similar way as is shown in Figure 14. The Lidar reference air-snow interface falls into areas which could be considered noise in the logarithmic scale intensity return while the algorithm was successful in identifying the first peak that satisfied the logarithmic threshold.

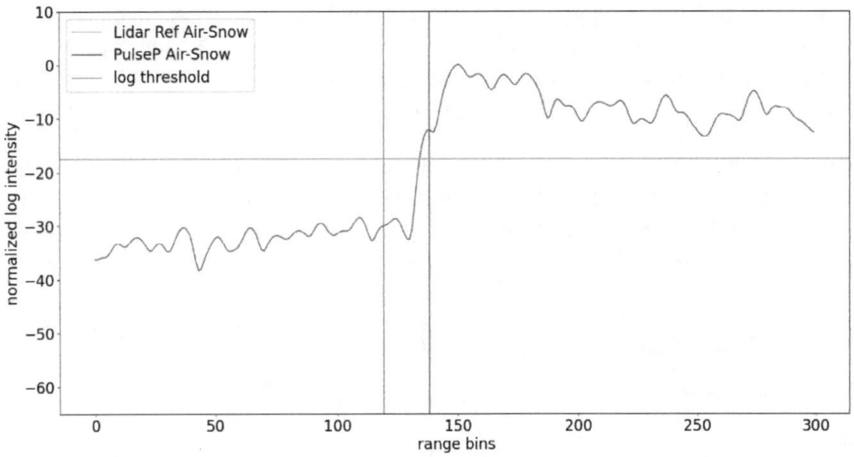

Figure 14. Sample underestimating waveforms for pulse peakiness where the Lidar reference air-snow interface falls below the logarithmic threshold.

As pulse peakiness left value alone normally cannot uniquely identify a peak as the air-snow interface, it is likely that before the interface peak, there would be numerous peaks with similar pulse peakiness left values. Consequently, it is extremely important to exclude as many non-interface peaks before the interface peak as possible to increase the chance of it being the first that exceeds the pulse peakiness left value threshold. This is achieved through the application of a logarithmic threshold. By lowering the threshold to accommodate the type of underestimation shown in Figure 14 the result would most likely lead to an overestimation of snow depth. An alternative is to apply a post algorithm bias correction based on this dataset.

To achieve this, the data were split in half into training and testing subsets, for which an average of underestimation was calculated for all the waveforms in the training subset. This average was used as the bias correction for all the waveforms in the testing set. Figure 15 shows the comparison between the un-corrected scatterplot and the corrected scatterplot. The average bias for the training set in this case is approximately -4 cm and the MAE after the correction decreased approximately 1 cm. The reason for such a small improvement is that the bias correction is applied to all the waveforms in the testing sets including those that are grossly overestimating. More intricate solutions to this bias correction, such as the identification and character generalization of underestimating waveforms should further improve the results.

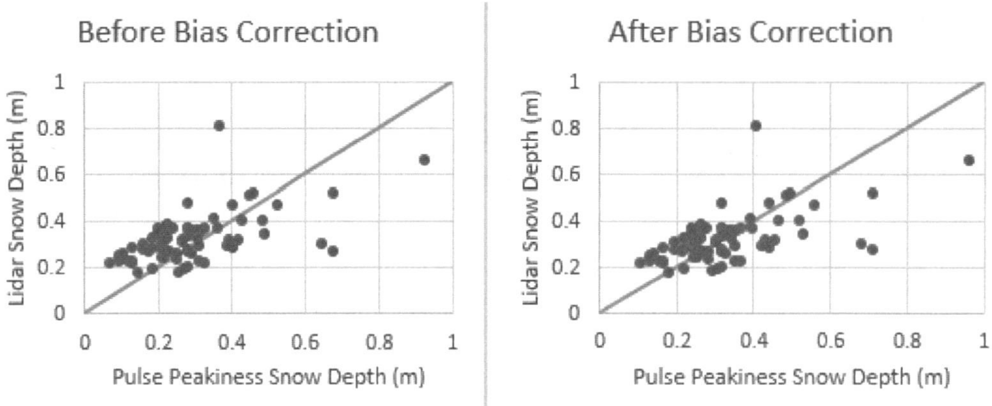

Figure 15. Before and after bias correction.

With respect to the overestimation problem, the main reason is the complexity of the waveform. A complex waveform here is defined as a waveform that has more than one peak that is close to the maximum return, in other words, a waveform that has no clear single maximum return. There can be a number of reasons why a waveform may have multiple "maximum" returns, but one plausible possibility is that a feature in the snowpack has a similar dielectric response to the ground surface.

This kind of waveform is inconsistent with the baseline assumption of a clear single maximum return peak being the snow-ground interface, and therefore disrupts the snow depth calculation which is the distance between air-snow peak and the maximum return. Figures 16 and 17 show examples of overestimation in dwarf shrubs and lichen.

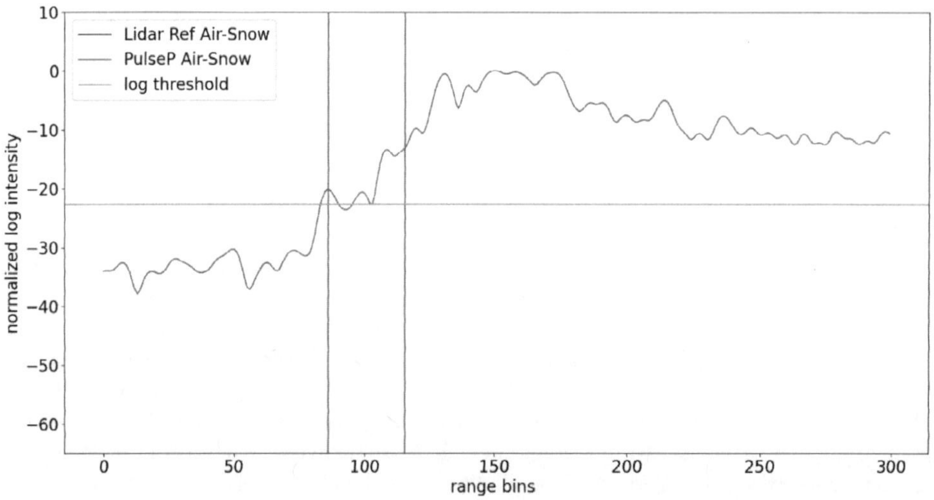

Figure 16. Sample overestimation for dwarf shrubs where there are multiple peaks close to the maximum peak.

To counteract this problem, one possible way is to apply a signal-to-noise ratio (SNR) threshold that excludes waveforms like the one shown in Figure 16. For such waveforms, adding another screening criterion that requires the waveform to have only 1 peak whose return is above a certain value should help filter them out. The challenge would be to set the correct threshold so that it does not filter out too many of the waveforms leaving the algorithm with insufficient valid outputs.

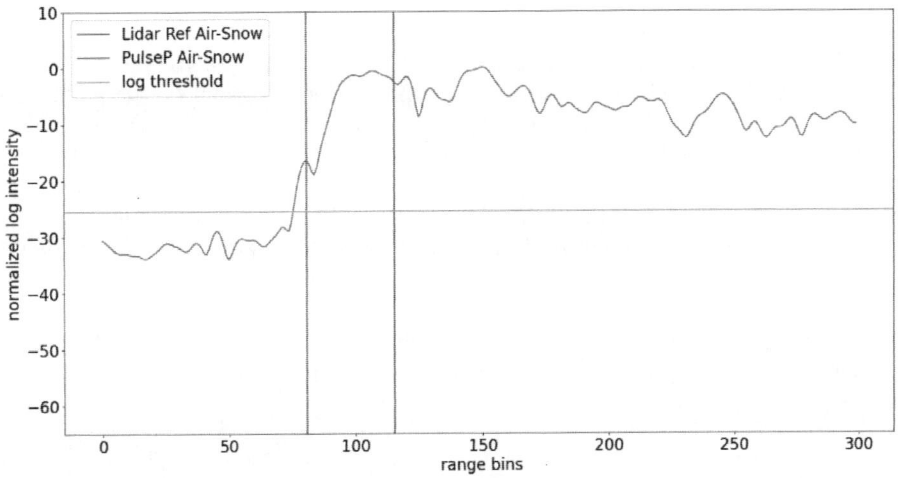

Figure 17. Sample overestimation for lichen where there are multiple peaks close to the maximum peak.

Error of Omission for the Pulse Peakiness Approach

Due to the limited number of valid outputs of the pulse peakiness approach, an analysis was done on the error of omission for dwarf shrub, tussock, and lichen. The reason for excluding trees, tall shrub and riparian shrub is that these landcover types have more complex waveform responses and therefore are omitted from the algorithm.

Table 5 shows the distribution statistics for Lidar points belonging to dwarf shrub (500), tussock (1224), and lichen (559) as well as those with a valid pulse peakiness output (135, 141, & 74). It demonstrates that for dwarf shrub and lichen, even though pulse peakiness filters excluded parts of the data, the distributions are still similar. However, for tussock, the pulse peakinesss distribution has shifted slightly to lower snow depth, which is further reinforced by the histogram shown in Figure 18. The pulse peakiness output is lacking representation of snow depth that is above 0.4m.

Table 5. Distribution statistics for Lidar snow depth and pulse peakiness valid output.

	Dwarf Shrub		Tussock		Lichen	
	Lidar	Pulse Peakiness	Lidar	Pulse Peakiness	Lidar	Pulse Peakiness
Mean (m)	0.32	0.33	0.38	0.28	0.44	0.37
Median (m)	0.30	0.28	0.36	0.25	0.42	0.34
Std (m)	0.14	0.19	0.14	0.13	0.16	0.16

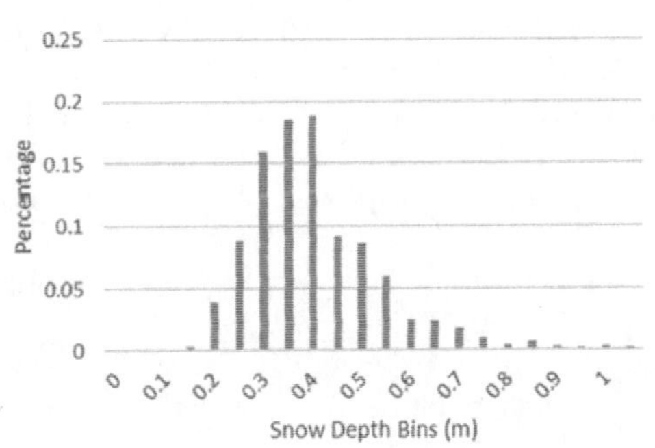

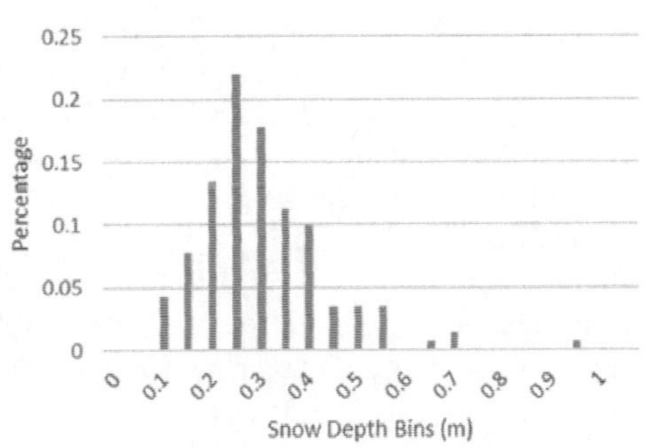

Figure 13. Distribution statistics for Lidar snow depth (left) and pulse peakiness (right) valid output.

CONCLUSION

The overall objective of this thesis is to test the applicability of a standard interface-based snow depth retrieval algorithms for snow on sea ice to snow on land. In principle, the pulse peakiness approach from Julita *et al.* (2021) should be valid for applicability to snow on land because the snow-ground interface provides a strong dielectric contrast along with the snow-air interface; these contrasts should be somewhat similar to snow on sea ice. However, the complex ground characteristics narrowed the locations where the pulse peakiness method could yield accurate results. More specifically, the presence of tall surface vegetation combined with the hummocky terrain meant that a considerably large percentage of the waveforms were discarded, and to some degree refocused the effort into how to filter out complex waveforms that are too ambiguous for the method. In conclusion, for snow on land, if the ground is generally a flat surface with minimal vegetation, it is expected that the interface-based pulse peakiness algorithm will have a good performance.

REFERENCES

Alfred Wegener Institute, Helmholtz Centre for Polar and Marine Research. 2019. *Awi als toolbox.* https://github.com/awi-als-toolbox/awi-als-toolbox. GitHub.

Anders K, Antonova S, Wegener A, Höfle B, Marsh P. 2018. Airborne laser scanning (als) point clouds of Trail Valley Creek, NWT, Canada (2016) vulnerability of tundra lakes to climate change view project 4d near real-time environmental monitoring view project. Retrieved from *https://www.uni-heidelberg.de/permasar. doi: 10.1594/PANGAEA.894884.*

Boike J, Grünberg I. 2019. Vegetation map of Trail Valley Creek, Northwest Territories, Canada. *doi: 10.1594/PANGAEA.904270.*

Croce P, Formichi P, Landi F, Mercogliano P, Bucchignani E, Dosio A, Dimova S. 2018. The snow load in Europe and the climate change. *Climate Risk Management*, **20**: 138-154. *doi: 10.1016/j.crm.2018.03.001.*

Cui Y, Xiong C, Shi J, Jiang L, Peng B, Ji D, Zhao T. 2016. Estimating snow water equivalent with backscattering at X and Ku bands. In: *2016 IEEE International Geoscience and Remote Sensing Symposium (IGARSS):* 3454-3457. *doi: 10.1109/IGARSS.2016.7729893.*

Geldsetzer T, Langlois A, Yackel J. 2009. Dielectric properties of brine-wetted snow on first-year sea ice. *Cold Regions Science and Technology,* **58**: 47-56. *doi: 10.1016/j.coldregions. 2009.03.009.*

Hendricks S, Ricker R, Jutila A. 2019. Icebird 2019 winter icesat-2 validation data acquisition report.

Jutila A, King J, Paden J, Ricker R, Hendricks S, Polashenski C, Helm V, Binder T, Haas C. 2021. High-resolution snow depth on arctic sea ice from low-altitude airborne microwave radar data. *IEEE Transactions on Geoscience and Remote Sensing*, **60**. *doi: 10.1109/TGRS.2021.3063756.*

King J, Toose P, Silis A, Derksen C. 2020. TVCSnow 2018-2019 tundra snow depth probe measurements. Retrieved from *https://doi.org/10.5281/zenodo.4021401.*

King J, Brady M, Newman T. 2020. kingjml/pysnowradar: Updated IEEE TGRS submission. Retrieved from *https://doi.org/10.5281/zenodo.4071947 doi: 10.5281/zenodo.4071947.*

Koenig LS, Ivanoff A, Alexander PM, MacGregor JA, Fettweis X, Panzer B, Paden JD, Forster RR, Das I, McConnell JR, Tedesco M, Leuschen C, Gogineni P. 2016. Annual Greenland accumulation rates (2009-2012) from airborne snow radar. *Cryosphere,* **10**: 1739-1752. *doi: 10.5194/tc-10-1739-2016.*

Kurtz NT, Farrell SL. 2011. Large-scale surveys of snow depth on arctic sea ice from operation icebridge. *Geophysical Research Letters*, **38**(20). *doi: 10.1029/2011GL049216.*

Kurtz NT, Farrell SL, Studinger M, Galin N, Harbeck JP, Lindsay R, Onana VD, Panzer B, Sonntag JG. 2013. Sea ice thickness, freeboard, and snow depth products from operation icebridge airborne data. *Cryosphere,* **7**: 1035-1056. *doi: 10.5194/tc-7-1035-2013.*

Kurtz NT, Galin N, Studinger M. 2014. An improved Cryosat-2 sea ice freeboard retrieval algorithm through the use of waveform fitting. *Cryosphere,* **8**: 1217-1237. *doi: 10.5194/tc-8-1217-2014.*

Kwok R, Maksym T. 2014. Snow depth of the Weddell and Bellingshausen sea ice covers from ice bridge surveys in 2010 and 2011: An examination. *Journal of Geophysical Research: Oceans,* **119**(7): 4141-4167. *doi: 10.1002/2014JC009943.*

Kwok R, Kurtz NT, Brucker L, Ivanoff A, Newman T, Farrell SL, King J, Howell S, Webster MA, Paden J, Leuschen C, MacGregor JA, Richter-Menge J, Harbeck J, Tschudi M. 2017. Intercomparison of snow depth retrievals over Arctic sea ice from radar data acquired by Operation IceBridge. *Cryosphere,* **11**: 2571-2593. *doi: 10.5194/tc-11-2571-2017.*

Leinss S, Wiesmann A, Lemmetyinen J, Hajnsek I. 2015. Snow water equivalent of dry snow measured by differential interferometry. *IEEE Journal of Selected Topics in Applied Earth Observations and Remote Sensing,* **8**: 3773-3790. *doi: 10.1109/JSTARS.2015.2432031.*

Marshall HP, Koh G. 2008. FMCW radars for snow research. *Cold Regions Science and Technology,* **52**: 118-131. *doi: 10.1016/j.coldregions.2007.04.008.*

Newman T, Farrell SL, Richter-Menge J, Connor LN, Kurtz NT, Elder BC, McAdoo D. 2014. Assessment of radar derived snow depth over Arctic sea ice. *Journal of Geophysical Research: Oceans,* **119**(12): 8578-8602. *doi: 10.1002/2014JC010284.*

Oveisgharan S, Esteban-Fernandez D, Waliser D, Friedl R, Nghiem S, Zeng X. 2020. Evaluating the preconditions of two remote sensing SWE retrieval algorithms over the US. *Remote Sensing,* **12**(12): 2021. *doi: 10.3390/rs12122021.*

Palmer MJ, Burn CR, Kokelj SV. 2012. Factors influencing permafrost temperatures across tree line in the uplands east of the Mackenzie delta, 2004-2010. *Canadian Journal of Earth Sciences,* **49**: 877-894. *doi: 10.1139/E2012-002.*

Pohl S, Davison B, Marsh P, Pietroniro A. 2005. Modelling spatially distributed snowmelt and meltwater runoff in a small Arctic catchment with a hydrology land-surface scheme (WATCLASS). *Atmosphere – Ocean,* **43**: 193-211. *doi: 10.3137/ao.430301.*

Pomeroy JW, Gray DM, Shook KR, Toth B, Essery RLH, Pietroniro A, Hedstrom N. 1998. An evaluation of snow accumulation and ablation processes for land surface modelling. *Hydrological Processes,* **12**(15): 2339-2367. *doi:10.1002/(SICI)1099-1085(199812)12: 15<2339::AID-HYP800>3.0.CO;2-L.*

SnowHydro. n.d. GPS snow depth probe. Retrieved from *http://www.snowhydro.com/products/column2.html*

Verret M, Wang Y, Bjornson J, Lacelle D. 2019. Hummocks in alpine tundra, northern British Columbia, Canada: Distribution, morphology and organic carbon composition. *Arctic Science,* **5**: 127-147. *doi: 10.1139/as-2018-0021.*

Yueh SH, Dinardo SJ, Akgiray A, West R, Cline DW, Elder K. 2009. Airborne Ku band polarimetric radar remote sensing of terrestrial snow cover. *IEEE Transactions on Geoscience and Remote Sensing,* **47**(10): 3347-3364. *doi: 10.1109/TGRS.2009.2022945.*

Enabling Low Latency Snow Pit Data

PUNEETH YOGANANDA[1], ROGER DE ROO[1], RUZBEH AKBAR[2], AND AGNELO SILVA[3]

ABSTRACT

Current methods of monitoring snow water equivalent (SWE) from space, such as measuring the differential scatter darkening in the radio brightness at 19 and 37 GHz, rely on snow microphysical properties, such as grain size, in addition to the snow macroscopic properties like snow depth. The algorithms to invert the observations to SWE or snow depth are region-specific and require substantial ground truth.

The research addresses this concern by automating the collection of snow ground truth. We will do this by integrating two technologies that already exist: 1. the SoilSCAPE system enables the collection of low-latency soil moisture data from areas that are large enough to be representative of a passive microwave footprint. 2. The University of Michigan snow sensor is a small, easily replicated device that logs snow's temperature, density, and grain size, moisture and ambient light levels. The bulk of the effort is on implementing the software needed allow the two systems to communicate, and thereby turn the SoilSCAPE system into one that can also monitor snow over a wide area.

Advantages of merging the two systems include giving to the snow sensor the intelligence of the SoilSCAPE system. Power management of a device embedded in the snowpack is important to preserving the snowpack properties. Judicious alterations of the measurement schedule, possible with the SoilSCAPE system, will enable rapid measurements of the snowpack when the snow changes, and sparse sampling of the snowpack when excess power dissipation is undesired.

The recent work includes successful data transmission of snow data from one snow sensor to the SoilSCAPE's End Device (ED) over UART, and the ED sending the data to a Local Coordinator (LC) wirelessly in a lab setting. The near future work involves expanding the network by connecting four snow sensors to the ED. This will enable space-time sampling of snow.

[1] Climate and Space Sciences and Engineering, University of Michigan, Ann Arbor, USA
[2] Department of Civil and Environmental Engineering, Massachusetts Institute of Technology, Cambridge, MA, USA
[3] METER Group Inc., Pullman, WA, USA

The Relationship Between Arctic Oscillation and Intense Cyclone Patterns and their Impacts on Summer Precipitation Distribution in the Canadian Arctic

XIAOMENG ZUO[1] AND LAURA BROWN[1]

ABSTRACT

A trend analysis of the Arctic System Reanalysis Version 2 (ASRv2) data from 2000 to 2015 showed that changes occurring with summer rainfall and snowfall are synchronized in the Canadian Arctic, rather than changing between forms. The precipitation in July showed a significant decreasing trend between 75° N and 80° N, while a significant increasing trend was found in September between 70° N and 80° N. Comparing the precipitation distribution in the Canadian Arctic with low-pressure system (center pressure < 1000 hpa) frequency and intensity patterns in the same study period showed high consistency, while the strength of the Arctic Oscillation (AO) also showed varying degrees of correlation with the frequency and intensity of the cyclones in the study area, thus affecting the distribution of precipitation. Visualizing the wind speed reanalysis data for each summer month showed that the distribution of intense cyclones is related to the positive or negative AO. The negative AO index in July in the 16 years corresponds to a southern movement of the jet stream, thus higher intensity cyclones, with local Laplacian over 15 mPa km^{-2}, are mostly observed in the areas around 60° N and further south, resulting in less precipitation in the northern high latitudes. On the contrary, the positive AO index in September corresponds to the northward shift in the jet stream, and more intense cyclones are observed between 70° N and 80° N, which brought more precipitation to the high latitude area in Canadian Arctic in the summertime. Meanwhile, strong positive AO index months are found to have more snow than negative AO index months in the higher latitudes. Finally, along the eastern coast of the Canadian Arctic, higher snow amounts, corresponding with more intense cyclones, are often found in positive AO index months, while this pattern is not evident for rain.

[1] Department of Geography, Geomatics and Environment, University of Toronto Mississauga, Mississauga, ON, Canada

Session #2: In situ Measurement of Snow and Solid Precipitation: Advances, Challenges, and Ongoing Issues

What if you put a Phone on a Drone?

FRASER KING[1], RICHARD KELLY[1], AND CHRISTOPHER FLETCHER[1]

ABSTRACT

Snow has significant impacts on springtime flooding, water resource management practices, and the global water-energy budget. *In situ* observations are considered some of the highest quality measurements of snow depth available and are useful constraints for climate model and reanalysis system estimates of snow water content. The application of laser altimetry (LiDAR-Light Detection and Ranging) for measuring snow depth has proven an effective method for quickly and accurately observing large areas, however this technique has traditionally been quite expensive to perform in practice due to the high cost of the necessary equipment coupled with required specialized training. In this work, we examine the capabilities of the iPhone 12 Pro LiDAR (iLiDAR) when attached to a consumer-grade DJI Phantom 4 drone (iDrone) in estimating snow depth at two study sites in southern Ontario, Canada. These sites were selected as representative locations of typical snow-covered areas including an open field with low-lying vegetation, as well as a sheltered site with a sloping surface and partial canopy cover. Initial comparisons between iDrone iLiDAR depth estimates and colocated snow ruler measurements demonstrate strong positive agreement with a mean squared error $<$ 4 cm, and low absolute mean bias of approximately 1 cm. Our results suggest that the intersection of these two technologies defines a novel, low-cost alternative to traditional LiDAR-based snow depth measurement systems, while maintaining a high degree of accuracy. Furthermore, the improved accessibility of this method provides new opportunities for citizen science-based snow depth contributions to local weather networks and in turn, the potential for improved regional forecasting of snow melt.

[1] Department of Geography and Environmental Management, University of Waterloo, Waterloo, ON, Canada

Transfer Function Development for Adjusting Precipitation Observations in Arctic and Maritime Climate Conditions

EVA MEKIS[1], AMBER ROSS[2], AND CRAIG SMITH[2]

ABSTRACT

The replacement of manual rain and snow observations with automated gauges is a difficult challenge for the user community and is leading to extensive research of these new instruments, installations, and measurements with the objective to quantify any underestimation and adjust the measurement bias. Key users, such as Numerical Weather Prediction (NWP) and Canadian Precipitation Analysis (CaPA) have identified the need to develop and apply transfer functions (TF) for precipitation gauge wind bias adjustments applicable for diverse Canadian climates. The WMO Solid Precipitation Inter-Comparison Experiment (SPICE) developed eight site-specific and one universal TF for the adjustment of solid precipitation measurements using data representing five different climate zones, namely Alpine, Northern Boreal, wet and dry Continental and Maritime climate. Transfer function development requires the availability of a Double-Fence Automated Reference (DFAR) installation for use as the reference precipitation observation in the estimation. Further to the three Canadian supersites that participated in the WMO-SPICE intercomparison experiment, two more locations in Canada with overlapping reference gauge observations are available. The Iqaluit supersite in Nunavut represents a typical Arctic climate with low snowfall amounts, cold temperatures, and blowing snow conditions, while the St John's site in Newfoundland has a maritime climate with high winds and frequent light and mixed precipitation conditions.

The focus of this presentation is to estimate the TF for all possible existing configurations [single Alter-shielded / unshielded / Belfort double Alter-shielded Geonor or Pluvio2 gauges] at Iqaluit and St John's sites. The major steps include 1) process and quality control all precipitation gauge, wind, temperature, and disdrometer data; 2) aggregate high frequency data to 30-minute and 60-minute intervals; 3) estimate transfer function parameters; and 4) validate the results. At the end, the results will be compared to the existing WMO-SPICE transfer functions and the possibility of applying climate dependent transfer functions across the Canadian networks will also be examined.

---

[1] Environment and Climate Change Canada, Toronto, ON, Canada
[2] Environment and Climate Change Canada, Saskatoon, SK, Canada

Snow Specific Surface Area: Margins of Error and Best Methods for the IceCube by A2 Photonic Sensors

KAITLIN MEYER[1] AND MICHAEL DURAND[1]

ABSTRACT

Snow specific surface area (SSA) is a minute physical property of snow that governs reflectance and plays an important role in microwave emission modeling and estimations of Earth's energy budget. While SSA is an important measurement, there are many challenges to acquiring *in situ* measurements. This project works with the IceCube, which measures snow SSA using an optical method in the field. The IceCube is highly desirable for its ability to make in field measurements, although there are some uncertainties about its margin of error. Some studies that have used the IceCube have noted a difference between IceCube SSA and micro-CT SSA up to 25% (Leppõnen *et al.*, 2018) and even higher in some cases. This project will take SSA samples using both IceCube and micro-CT from a wide variety of snow packs. Results will be compared and analyzed with the goal of laying out potential sources of error, identifying trends in the SSA differences, giving a better margin of error for the IceCube, and prescribing the best methods for sampling with the IceCube. We hope that this will be useful to those who are not experts in microstructure but need accurate SSA measurements to validate modeling products, remote sensing data, as well as for other areas of study.

[1] School of Earth Sciences, The Ohio State University, Columbus, OH, USA

Traceability and Catch-Efficiency of the Lambrecht rain[e]H3 Automated Precipitation Gauge for Measuring Precipitation in Canadian Operational Networks

AMBER ROSS[1], EVA MEKIS[2], AND CRAIG SMITH[1]

ABSTRACT

As organizations are looking to streamline observation networks, many are transitioning from manual observations to automated gauges for the measurement of precipitation. In order to help ensure homogeneity, intercomparisons with reference methods and gauges are acutely important. These studies ultimately lead to better precipitation data with fewer errors and biases. The key condition of providing these intercomparisons is the availability of overlapping data between the old and new systems or with reference precipitation observations along with the continuous support and maintenance of these intercomparison research sites. Since climate research and models rely on accurate meteorological measurements, it is important to know the limitations of gauges and methods used to collect data.

The aviation weather monitoring network operated by NAV CANADA, in an effort to modernize, is installing Lambrecht rain[e]H3 heated tipping bucket / weighing type gauges to replace manual (Nipher or Type B gauges or snow ruler) observations of precipitation. This has implications on the accuracy of reporting precipitation, especially snow events which are more substantially prone to measurement errors. Collaboration between NAV CANADA and Environment and Climate Change Canada (ECCC) has provided an opportunity to analyze existing overlapping manual and automated measurements. The overall intercomparison study includes four sites across Canada: Downsview (ON), Prince George (BC), Whitehorse (YT), and Dorval (QC). Additionally, overlapping observations of the Lambrecht gauge and DFAR (Double Fence Automated Reference) measurements are available for the 2021-22 winter season at the ECCC Bratt's Lake (SK) supersite. Preliminary results of two setups from Downsview and Bratt's Lake show that wind speed and precipitation phase greatly impact the Lambrecht gauge catch efficiency.

INTRODUCTION

Errors in solid precipitation measurements due to gauge configuration and wind speed are well documented (e.g. Sevruk *et al.*, 1991; Goodison *et al.*, 1998; Kochendorfer *et al.*, 2017a). The systematic bias due to wind induced undercatch can lead to the underestimation of solid precipitation and is an important consideration when selecting and deploying new automated precipitation gauges and accompanying wind shields to cold regions.

Measurement intercomparisons using overlapping observations are crucial for long term data homogeneity, especially given that manual observation networks are in decline across Canada (Mekis *et al.*, 2018). As an example, overlapping observations were utilized in the development of

[1] Environment and Climate Change Canada, Saskatoon, SK, Canada
[2] Environment and Climate Change Canada, Toronto, ON, Canada

the Adjusted and Homogenized Canadian Climate Dataset (AHCCD) precipitation time series (Vincent and Mekis, 2009; Mekis and Vincent, 2011). For many AHCCD stations, it was necessary to merge observations from nearby stations to produce longer time series of rainfall and snowfall for trend analysis. Standardized ratios comparison procedures of the overlapping observations were used to determine if changes in precipitation observations created any artificial discontinuities at the joining dates. Similarly, to ensure data continuity, adjustments may be required when switching to new precipitation gauge configurations, making the availability of overlapping reference observations essential.

This extended abstract compares precipitation measurements from the Lambrecht rain[e]H3 heated tipping bucket / weighing gauge to reference methods and gauges from the ECCC Bratt's Lake supersite and the ECCC Downsview site. The overall project objective is twofold: 1) setup a possible transfer function between the old manual to the new automated systems observations and 2) assess the accuracy and wind bias of the Lambrecht gauge using the reference DFAR installation. Some preliminary results are indicated here for each.

DATA AND METHODOLOGY

The single Alter (SA) shielded Lambrecht gauge was installed at both the Downsview and Bratt's Lake locations. The Lambrecht installation at Bratt's Lake is shown in Figure 1. First the 20-second Lambrecht data obtained from the data logger were converted to 1-minute intervals by taking the sum of the non-missing values in each 1-minute time interval. The 1-minute data were further aggregated to match the reference observation periods of each site (described below). Quality control (QC) included the removal of duplicate time stamps and the creation of equidistant time series with no missing time stamps. It also included the filtering out of periods with missing values as well as periods where an error was reported in the Lambrecht data such as heating errors or temperature sensor errors. Additionally, visual examination of the Lambrecht gauge time series was completed for further erroneous data identification.

Figure 1. Single Alter shielded Lambrecht rain[e]H3 heated tipping bucket / weighing gauge installed at the Bratt's Lake (SK) supersite.

The reference at the Bratt's Lake supersite is the internationally recognized World Meteorological Organization Solid Precipitation Inter-Comparison Experiment (WMO-SPICE) DFAR with a Pluvio^2 automated gauge in the middle. The study period is Oct 2021 – Apr 2022. Precipitation and ancillary 1-minute data were downloaded directly from site data loggers and filed into equidistant monthly files with no missing time stamps. The DFAR precipitation data were quality controlled through an automated process which removed out-of-range outliers and data jumps related to gauge servicing (bucket emptying and/or charging). This was followed by manual inspection and removal of any erroneous data missed by the automated QC process. The DFAR cumulative time series was then filtered using the Supervised Neutral Aggregating Filter (NAF-S) to adjust for evaporation and other spurious data (Ross *et al.*, 2020). DFAR precipitation was accummulated and wind speed and temperature were averaged over 24-hour periods. The 1-minute Lambrecht data were summed to match the 24-hour periods selected for comparison. Precipitation phase was determined using WMO-SPICE air temperature (T_{air}) thresholds where snow is defined as T_{air} < -2°C, mixed precipitation as -2°C ≤ T_{air} ≤ 2°C, and rain as T_{air} > 2°C (Kochendorfer *et al.*, 2017b).

The intercomparison at the Downsview site involved manual reference observations as is described below. The comparison was completed for the longer Nov 2019 – Jun 2021 period. Manual observations were obtained from the web-based COOLTAP (Cooperative Online Temperature and Precipitation) reporting system. Observations from COOLTAP include maximum and minimum temperatures, rainfall, snowfall, and snow depth data taken once or twice daily. Data entered by users are first stored and then synchronized in the COOLTAP database every few minutes. Given that manual precipitation observations can be made with various methodologies (i.e. Nipher or Type B gauge or snow ruler) and reported as rainfall, snowfall, or SWE (Snowfall Water Equivalent) and that our objective is to compare total precipitation accumulations, Table 1 outlines the procedure used to calculate Total Precipitation from the reported manual observations. For example, Case 3 describes that when COOLTAP reports Snowfall > 0 cm, SWE > 0 mm, and Rainfall = 0 mm, then Total Precipitation is derived from the COOLTAP SWE column. COOLTAP quality control included filing the data into a non-equidistant time series based on observation date/time, the removal of duplicate time stamps, and time zone conversion from local to UTC. In most cases, manual observations were conducted at approximately 8 am and 4 pm local time creating uneven precipitation accumulation periods of approximately 8 and 16 hours. Overlapping Lambrecht (1-minute) and ancillary data were then aggregated to these reference observation periods with temperature and wind speed averaged and Lambrecht data summed. For precision, the initial Lambrecht data was aggregated to the minute instead of the hour because some manual observations were conducted 15 to 30 minutes from the top of the hour. Precipitation phase was derived through the COOLTAP 'remarks' column where available, and via the air temperature thresholds described earlier when human precipitation type observations were not available.

Table 1. Downsview COOLTAP manual observation procedure applied for deriving total precipitation. Rainfall was measured using a Type B gauge, snowfall was measured using a snow ruler, and Snowfall Water Equivalent (SWE) was derived from either a Type B or Nipher gauge.

	Downsview procedure for deriving total precipitation			
Case	Rainfall (mm)	Snowfall (cm)	SWE (mm)	Total Precipitation (mm)
1	0	0	0	0
2	0	0	>0	SWE
3	0	>0	>0	SWE
4	>0	0	0	rainfall
5	>0	>0	0	rainfall
6	0	>0	0	10 x snowfall
7	>0	0	>0	rainfall + SWE
8	>0	>0	>0	rainfall + SWE

RESULTS

The accumulated precipitation observation time series for each site was plotted for visual assessment of the accuracy and effectiveness of the Lambrecht gauge performance when compared to the respective references (Figure 2). In addition, statistical measures were calculated including 1) total number of events for each precipitation phase; 2) probability of detection (% of events where both Lambrecht > 0 & reference > 0); 3) false alarm reporting (% of events where Lambrecht > 0 & reference = 0); 4) relative total catch (total accumulated Lambrecht precipitation compared to the reference expressed as %); 5) root mean square deviation (RMSD) normalized to the range of reference observations (maximum reference observation minus minimum reference observation);

and 6) correlation coefficient (r) between the Lambrecht gauge and the reference (Tables 2 & 3). The statistics excluded any observation periods where the reference and Lambrecht gauge both reported 0 mm. The climate characteristics of the two tested locations were quite different. For precipitation events (all phases combined), the mean wind speed was 5 m s^{-1} at Bratt's Lake and 1 m s^{-1} at Downsview, while the mean air temperature was -11.7 °C at Bratt's Lake and 7.4 °C at Downsview.

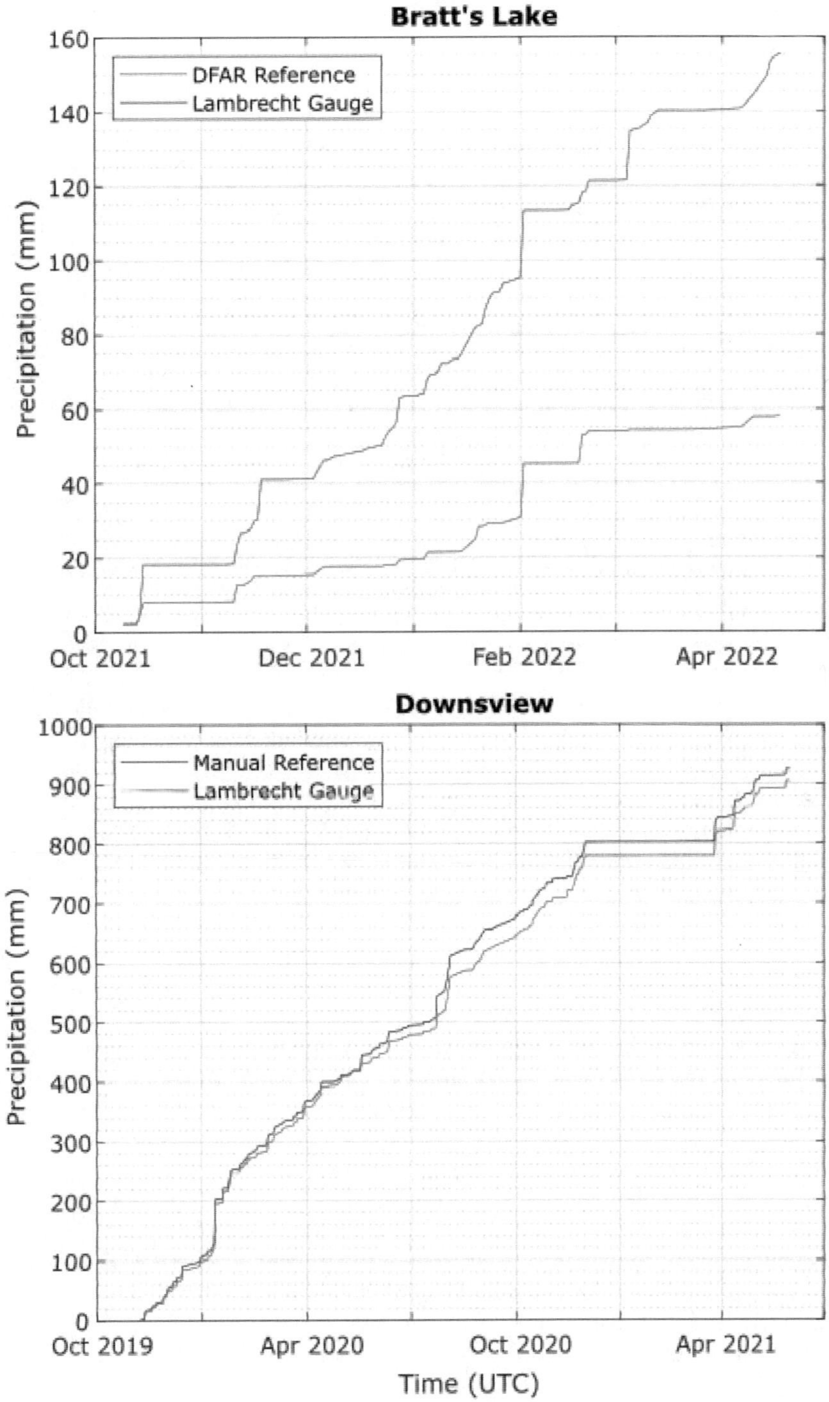

Figure 2. Accumulated precipitation observation time series for all precipitation phases at Bratt's Lake (top) and Downsview (bottom). Note that data from Bratt's Lake is only available for the winter of 2021-22.

When comparing the Lambrecht gauge to the reference WMO-SPICE DFAR at Bratt's Lake, the accumulated winter precipitation time series shows considerable differences. This underlines the overall undercatch of the Lambrecht gauge compared to the reference under cold and windy conditions. In contrast, the Lambrecht gauge performs comparatively well to the manual reference at Downsview with the two time series tracking closely to each other.

Table 2. Bratt's Lake daily [24 hour] statistics for rain, snow, mixed, and all precipitation phases.

	Bratt's Lake statistics					
	Total Number of Events	Probability of Detection (%)	False Alarm Reporting (%)	Relative Total Catch (%)	Normalized RMSD (mm)	*r*
Rain	7	100	14	88.4	0.100	0.97
Snow	72	44	38	32.1	0.145	0.64
Mixed	12	67	60	31.7	0.217	0.97
All	91	51	39	37.4	0.141	0.65

Table 3. Downsview 8 and 16 hour statistics for rain, snow, mixed, and all precipitation phases.

	Downsview statistics					
	Total Number of Events	Probability of Detection (%)	False Alarm Reporting (%)	Relative Total Catch (%)	Normalized RMSD (mm)	*r*
Rain	182	94	16	98.1	0.044	0.95
Snow	67	87	34	93.8	0.063	0.93
Mixed	22	100	36	101.5	0.045	0.99
All	271	93	22	97.7	0.040	0.95

At Bratt's Lake, relative total catch was moderately high for rain at 88% but very low for snow and mixed precipitation at around 32%. The probability of detection was perfect at 100% for rain (although the number of rain events is low during the study period) and only 44% for snow. False alarm reporting was the highest at 60% for mixed precipitation, followed by 38% for snow, and 14% for rain. The correlation coefficient (*r*) is high for rain and mixed events at 0.97 but relatively low for snow at 0.64. There were also a large number of snow events not detected by the Lambrecht gauge (i.e. the catch efficiency was 0). Out of the 72 snow events, 32 had the Lambrecht catch efficiency equal to 0. The DFAR reported 45.7 mm of precipitation (out of the total accumulation of 155.4 mm) for these 32 events where the Lambrecht reported 0 mm. This amount represents 29% of the total DFAR precipitation.

At Downsview, the relative total catch was very high (98% for rain, 94% for snow, and 102% for mixed precipitation). Probability of detection was also high for all precipitation phases with the lowest at 87% for snow. Similarly, snow had the lowest correlation coefficient at 0.93. False alarm reporting was 16% for rain, 34% for snow, and 36% for mixed precipitation. At Downsview, 2/3 of the cases were rain events while the number of mixed events (within the -2 °C and +2 °C range) were comparatively low (22 out of the total 271 events). The normalized RMSD for each precipitation phase was lower at Downsview when compared to Bratt's Lake.

The use of two different references is an important consideration when interpreting these results. When the manual Nipher gauge is used for comparison of the snow events at Downsview, its catch efficiency has the potential to be more influenced by wind as compared to the expectancy for a

DFAR configuration. However, because of the low wind speeds observed during this experiment at Downsview, the catch efficiency of the Nipher will be quite high and close to 100% (see Goodison, 1978). If the Nipher gauge would have been used in a similar fashion at Bratt's Lake (rather than the DFAR), its performance may not be as good because wind speeds are considerably higher at this location.

CONCLUSIONS

The present study compares the Lambrecht rain[e]H3 gauge performance to manual observations at Downsview (ON) and to DFAR reference observations at Bratt's Lake (SK). It was found that the Lambrecht gauge performed relatively well at Downsview but poorly at Bratt's Lake when compared to the respective references. The Lambrecht measurements correlate well to the manual observations at Downsview for all precipitation phases. At Bratt's Lake, significant deviations exist between the Lambrecht and DFAR observations with considerable undercatch for snow and mixed precipitation.

The two results obtained above are not quite comparable, since the reference used in the studies are different, but a large contributor to the difference in performance is the contrasting climate conditions at the two sites. While Bratt's Lake typically has lower temperatures, higher wind speeds, and more snow, Downsview has higher temperatures, lower wind speeds, and more rain events. The greatest determining factors in catch efficiency are shown to be wind speed and precipitation phase. The Lambrecht gauge performed well for rain events with low wind speeds (1 m s^{-1}) at Downsview, while higher wind speeds (5 m s^{-1}) combined with snow and mixed precipitation at Bratt's Lake are shown to cause notable undercatch.

It is found that transitioning from manual precipitation observations to the automated Lambrecht rain[e]H3 gauge in operational networks will cause uncertainty and inhomogeneity on long term climate time series. This will presumably cause a decline in total precipitation amounts due to a higher probability of undercatch if the gauge is deployed in cold regions with high wind speeds. It is anticipated that the impact will be less pronounced in warmer regions with fewer snow events and under lower wind speed conditions. Next steps include intercomparisons of overlapping Lambrecht and reference precipitation observations at the Prince George (BC), Whitehorse (YT), and Dorval (QC) sites. Analyzing further climate conditions to assess the Lambrecht gauge catch efficiency and develop transfer functions from manual observations to the new automated Lambrecht gauge will help provide further details and ensure data continuity and homogeneity moving forward.

REFERENCES

Goodison B. 1978. Accuracy of Canadian snow gauge measurements. *Journal of Applied Meteorology,* **17**(10): 1542-1548. *https://doi.org/10.1175/1520-0450(1978)017<1542:AOCSGM>2.0.CO;2*

Goodison B, Louie P, Yang D. 1998. The WMO solid precipitation measurement intercomparison. *World Meteorological Organization-Publications-WMO/TD,* **872**: 212.

Kochendorfer J, Rasmussen R, Wolff M, Baker B, Hall ME, Meyers T, Landolt S, Jachcik A, Isaksen K, Brækkan R, Leeper R. 2017a. The quantification and correction of wind-induced precipitation measurement errors. *Hydrology and Earth System Science,* **21**: 1973-1989. *https://doi.org/10.5194/hess-21-1973-2017*

Kochendorfer J, Nitu R, Wolff M, Mekis E, Rasmussen R, Baker B, Earle ME, Reverdin A, Wong K, Smith CD, Yang D, Roulet Y-A, Buisan S, Laine T, Lee G, Aceituno JLC, Alastrué J, Isaksen K, Meyers T, Brækkan R, Landolt S, Jachcik A, Poikonen A. 2017b. Analysis of single-Alter-shielded and unshielded measurements of mixed and solid precipitation from WMO-SPICE. *Hydrology and Earth System Sciences* **21**(7): 3525-3542. *https://doi.org/10.5194/hess-21-3525-2017*

Mekis E, Vincent LA. 2011. An overview of the second generation adjusted daily precipitation dataset for trend analysis in Canada. *Atmosphere-Ocean,* **49**(2): 163-177. *https://doi.org/10.1080/07055900.2011.583910*

Mekis E, Donaldson N, Reid J, Hoover J, Zucconi A, Li Q, Nitu R, Melo S. 2018. An overview of the surface-based precipitation observations at Environment and Climate Change Canada. *Atmosphere-Ocean,* **56**(2): 71-95. *https://doi.org/10.1080/07055900.2018.1433627*

Ross A, Smith CD, Barr A. 2020. An improved post-processing technique for automatic precipitation gauge time series. *Atmospheric Measurement Techniques,* **13**: 2979-2994. *https://doi.org/10.5194/amt-13-2979-2020*

Sevruk B, Hertig J-A, Spiess R. 1991. The effect of precipitation gauge orifice rim on the wind field deformation as investigated in a wind tunnel. *Atmosphere Environment,* **25A**: 1173-1179.

Vincent LA, Mekis E. 2009. Discontinuities due to joining precipitation station observations in Canada. *Journal of Applied Meteorology and Climatology,* **48**(1): 156-166. *https://doi.org/10.1175/2008JAMC2031.1*

Assessing the Impact of Transition from Single Alter-Shielded Geonor T-200B to Double Alter-Shielded Pluvio^{2}L Gauges on Winter Precipitation Measurements in the ECCC Operational Network

CRAIG SMITH[1], AMBER ROSS[1], AND EVA MEKIS[2]

ABSTRACT

As a key component of network modernization, Environment and Climate Change Canada (ECCC) is replacing single Alter-shielded Geonor T-200B (Geonor-SA) precipitation gauges with double Alter-shielded OTT Pluvio2L (P2L-DA) precipitation gauges. Geonor-SA gauges have been used in the ECCC climate network since the early 2000s. The replacement program with the P2L-DA configuration began as early as 2016. The number of P2L-DA gauges in the ECCC climate network was approximately 215 at the end of 2021. Following years of precipitation gauge intercomparisons, it is well recognized that more extensive shielding (i.e. double vs. single Alter-shields) results in a higher gauge catch efficiency (CE), especially for solid precipitation, increasing the probability of detection of light events and decreasing the measurement uncertainty. However, introducing this configuration into the network creates a measurement inhomogeneity, especially at cold and windy sites where the double shield will substantially increase the relative amount of observed winter precipitation as compared to the previously installed single shield. A universal transfer function (UTF) to adjust wind bias is currently available for single Alter-shielded gauges including the Geonor-SA configuration and has been applied to ECCC network measurements, with adjusted data published in the Government of Canada Open Data Portal. Unfortunately, a similar universal transfer function does not yet exist for double Alter-shielded gauges such as the P2L-DA configuration and therefore those measurements are not adjusted.

This analysis assesses the impact of two scenarios that may result in precipitation time series inhomogeneity: 1) transition from the unadjusted Geonor-SA to the unadjusted P2L-DA (realized by using the observations from the ECCC public archive); and 2) transition from the UTF adjusted Geonor-SA to the unadjusted P2L-DA (realized by using the current version of the published bias-adjusted data). The implications and solutions for these scenarios are explored using intercomparison data from the Bratt's Lake (SK) supersite, including overlapping data with the official ECCC climate station.

[1] Environment and Climate Change Canada, Saskatoon, SK, Canada
[2] Environment and Climate Change Canada, Toronto, ON, Canada

Session #3: Monitoring and Modeling Cryospheric Change

Spatial Variability of Lake Ice Thickness and Phenology on Sub-Arctic Lakes in Yellowknife, Northwest Territories

GIFTY ATTIAH[1], HOMA KHEYROLLAH POUR[1], AND ANDREA SCOTT[2]

ABSTRACT

Lake ice is a valuable resource to northern communities such as those within the Northwest Territories (NT). Ice roads are constructed on lake ice (e.g., the ice longest road in Yellowknife, NT, spreading over 80 lakes) during winter to haul goods to and from industrial establishments (e.g., mines) and for travel within communities. A major detriment to the ongoing use of ice roads are the changes in duration and thickness of lake ice due to climate warming. Studies show that a one-degree increase or decrease in air temperature leads to a 6-day almost linear change in ice cover duration making it an essential climate variable to monitor. Crucial knowledge on lake ice and temperature is however limited especially in sub-artic lakes due to logistical difficulties in collecting direct measurements. Methods adopted such as one-dimensional lake ice models rely on weather stations/*in situ* data as inputs which are sparse in the NT hence overlook several lakes. To address this limitation, this study uses remote sensing coupled with a thermodynamic lake ice model to monitor the spatial distribution, duration, and thickness of lake ice.

Over 500 small to medium lakes are monitored in this study. To effectively simulate lake ice thickness, an algorithm-based surface temperature for each lake was derived from the thermal bands of Landsat which showed good agreement with *in situ* data (1.88 °C > RMSE > 1.54 °C). Lake surface temperatures generated in addition to variables (wind speed (m s^{-1}), relative humidity (%), snowfall (m), and cloud cover (0-1)) and reanalysis (ERA5) data served as major inputs in the spatially-distributed thermodynamic model applied. Field work was conducted to collect ice thickness measurements across 10 lake sites in Yellowknife to evaluate model output. Output derived from model simulations not only demonstrate changes in lake ice thickness, but also highlights the spatial variability within lakes.

[1] Department of Geography and Environmental Studies, Wilfred Laurier University, Waterloo, ON, Canada
[2] Department of Systems Design Engineering, University of Waterloo, Waterloo, ON, Canada

Diagnosing the Future Hydrology of a Central Asian Glacierized Basin using a Hydrological-Glaciological Land Surface Model

OKAN AYGÜN[1], JOHN W. POMEROY[1], MARTYN P. CLARK[1], AND ALAIN PIETRONIRO[2]

ABSTRACT

Water resources in high mountain central Asia are strongly dependent on snow and glacier melt which are vulnerable to climate change. How climate change impacts hydrology has not previously been diagnosed in the region using physically-based glacio-hydrological models that can resolve mountain snow and glacier hydrology with confidence. This study assesses the impacts of projected climate change on the hydrology of the snow-covered Ala-Archa River Basin in the Tien Shan Mountains of Kyrgyzstan, a mountain headwater basin, 15% covered by glaciers. Current and future hydrological processes and streamflow were diagnosed using MESH, a Canadian physically-based hydrological land surface scheme with comprehensive representation of cold regions processes. The MESH Ala-Archa model was forced by the EM-Earth (0.1°) and ERA-5 (0.25°) datasets. Temperature and precipitation forcing data over 1991-2010 were seasonally perturbed using outcomes of CMIP5-AR5 subset for RCP 8.5 for the region over the 2081-2100 period. MESH streamflow prediction was good, with Kling-Gupta Efficiency (KGE) scores above 0.9 and percent bias below ±1% over both calibration and validation periods. Under the climate change scenarios examined, the snowfall ratio declined from 49% to 33%, snow-covered period decreased by two months, and peak SWE declined by 15%. Ice melt and snowmelt presently form 40% and 25% of the total runoff, respectively, but snowmelt became the largest source of runoff with warming and deglaciation. The timing of peak streamflow advanced from mid-July to early June, and both peak discharge and annual streamflow volume decreased by more than 50% for the 2081-2100 period compared to recent conditions. These results underline the need for renewed diagnostic assessments of water supply in high mountain headwaters of Central Asia to inform adaptation to climate change.

[1] Centre for Hydrology, University of Saskatchewan, Saskatoon, SK, Canada
[2] Department of Civil Engineering, University of Calgary, Calgary, AB, Canada

Investigating the Impact of Snow Cover on Permafrost Soil Temperatures in Modern Reanalysis and Data Assimilation Systems

TYLER HERRINGTON[1], CHRISTOPHER FLECTCHER[1], AND ANDRE ERLER[2]

ABSTRACT

Soil temperatures are required as input for hydrological models and for numerical weather prediction. At high latitudes, accurate permafrost representation is important as soil respiration from melting permafrost may act as a positive feedback on warming. Validation of Arctic soil temperatures in reanalysis and Land Data Assimilation System (LDAS) products (hereafter reanalysis products) has been historically limited because widespread *in situ* reference observations have generally been unavailable. Here we validate pan-Arctic soil temperatures from eight reanalysis and Land Data Assimilation System (LDAS) products, at 1-degree spatial resolution, using *in situ* soil temperature data from diverse measurement networks across Eurasia and North America. We find that most products are biased cold by 2-7 K across the Arctic. Near-surface soil temperature biases and Root Mean Square Error (RMSE) were generally largest in the cold season, and many products overestimate the observed variability in soil temperatures over the cold season. In addition, preliminary results show that the cold season RMSE in many products was more than 1.5 times as large when snow was present in satellite-based snow cover datasets, though there is substantial variability between products. We attempt to explain the large spread in cold season Arctic soil temperatures by reconciling differences in snow cover between reanalysis products and satellite-based snow cover, and variability in snow cover between reanalysis products. Our hypothesis is that RMSE and bias in soil temperature will be largest when reanalysis products and satellite snow-cover show substantial disagreement; likely when monthly mean air temperatures are close to the freezing point. We also examine a subset of reanalysis products at a higher resolution (0.05 degrees) to test the impact of spatial resolution on soil temperature performance, and preliminary results suggest that small improvements in pan-Arctic soil temperature biases may be achieved by using higher resolution soil temperature data.

[1] Department of Geography and Environmental Management, University of Waterloo, Waterloo, ON, Canada
[2] Aquanty, Waterloo, ON, Canada

Cryospheric Observation and Modeling for Improved Adaptation in Central Asia (CROMO-ADAPT)

MARTIN HOELZLE[1], MARTINA BARANDUN[2], CHRISTIAN HAUCK[1], CHRISTIN HILBICH[1], JOEL FIDDES[3], TAMARA MATHYS[1], ENRICO MATTEA[1], RODICA NITU[4], JEANNETTE NOETZLI[3], ERIC POHL[1], AND TOMAS SAKS[1]

ABSTRACT

Climate change is a challenge for humanity and the resulting global consequences will influence and endanger economies and livelihoods of coming generations. Strategies to improve climate resilience through adaptation must be based on sound baseline information, such as climate observations, especially on the Essential Climate Variables (ECVs) defined by the Global Climate Observing System (GCOS). Significant gaps exist however in the global climate observing system, particularly in High Mountain Asia. This data scarcity results in high uncertainties about the experienced as well as expected consequences of climate change, preventing sound anticipation of future developments and planning corresponding adaptation measures. A region where climate change is projected to have major impacts on water availability and occurrence of natural hazards is Central Asia's mountain regions.

A new project, named Cryospheric Observation and Modelling for improved Adaptation in Central Asia (CROMO-ADAPT), will address this gap by strengthening and (re-)establishing cryospheric (snow, glaciers, and permafrost) *in situ* monitoring systems in Central Asia. This project continues the improvements of glacier monitoring activities that have been ongoing for more than a decade by complementing it with snow and permafrost monitoring systems and corresponding capacity building in Central Asia. Ultimately, this project aims to develop climate resilience through climate information services and adaptation measures based on sound climate observations. Four watersheds, particularly in the headwater regions of Kyrgyzstan and Tajikistan are selected for a combined cryospheric observation strategy following the guidelines provided by the Global Cryosphere Watch (GCW) of the World Meteorological Organization (WMO) and the Global Framework for Climate Services (GFCS). This includes mainly the application of old and new *in situ* technology of snow, glacier, and permafrost long-term observations.

[1] Department of Geoscience, University of Fribourg, Fribourg, Switzerland
[2] Institute of Earth Observation, Bolzano, Italy
[3] WSL Institute for Snow and Avalanche Research SLF, Davos, Switzerland
4 World Meteorological Organization, Geneva, Switzerland

Comparison and Calibration of a Temperature-Index based Coupled Glacier Mass Balance Dynamic Evolution Model and Applications in Regional Melt Projections

AMANDA KOTILA[1], ANDREW BUSH[1], AND MONIREH FARAMARZI[1]

ABSTRACT

Projection of the changes in the dynamics of glacier accumulation, ablation, and melt runoff is crucial for the management and planning of downstream water use and aquatic ecosystems. The temperature-index models (TIMs) and energy-balanced approaches (EBMs) are widely used to predict glacier changes. In most studies, the TIMs and EBMs are applied with limited or no calibration of model parameters for regional estimates of glacier changes. Hence, they do not account for the spatially variable glacier characteristics for individual glaciers and a group of glaciers at a regional scale. On the other hand, the scarcity of high-resolution measured data at the scale of individual glaciers poses a significant challenge for predicting glacier changes at a regional scale.

This research compared the performance and predicted the uncertainty of a coupled glacier mass balance dynamic evolution model (CGMBDEM) using two TIMs for application in regional studies. The Classical Temperature Index Model (CTIM) and the Pellicciotti Temperature Index Model (PTIM) were examined using the best available data from Athabasca Glacier in Western Canada. The CTIM calculates melt using degree-day factors and a threshold temperature, whereas the PTIM incorporates melt due to radiative forcing. The CGMBDEM was calibrated, and an uncertainty prediction was made based on the optimized range of physical and meteorological parameters via Latin Hypercube Sampling and 95 Percent Prediction Uncertainty (95PPU). Calibration and uncertainty assessment results indicate that the CGMBDEM simulations using the CTIM tend to have a statistically better (bR2 = 0.726, p-factor = 0.77, r-factor = 1.04) performance than those using the PTIM (bR2 = 0.551, p-factor = 0.31, r-factor = 1.02). This demonstrates the sensitivity of model performance to TIM choice and associated parameters. We propose that the 95PPU based on optimized parameter ranges in our CTIM-based CGMBDEM can be useful in modeling regional changes in glacier melt runoff, mass-balance, and evolution while balancing errors due to the issue of parameter transferability.

[1] Department of Earth and Atmospheric Sciences, University of Alberta, Edmonton, AB, Canada

Changing Arctic Snow and Ice: *In situ* Monitoring using Digital Imagery

BRIANNA LANE[1] AND LAURA BROWN[1]

ABSTRACT

In the Arctic region, a rapid loss of snow and ice cover has been observed in recent years. The effects of this loss have significant impacts on global climate, hydrological events, biological processes, and human populations. Previous studies have measured changes in snow and ice cover using satellite imagery and model-based approaches. Novel research is beginning to utilize ground-based automated camera systems for *in situ* monitoring of lake and river ice. The overall aim of this study is to maximize the usage of field imagery for snow and ice studies within the context of the changing Arctic climate. This study is focused on monitoring lake ice and snow at five study lakes (Resolute Lake, Small Lake, Plateau Lake, North Lake, and Hunting Camp Lake) near Resolute and Nanuit Itillinga, Nunavut in the central Canadian High Arctic with data available from 2016-2021. The objective of the research is to develop a feasible method for snow and lake ice data quantification from *in situ* digital imagery and examine the recent variability in snow and ice phenology using the method developed. Preliminary results using image segmentation techniques show promising results for tracking the progression of ice formation and snow redistribution across the lakes. Ongoing work is focused on assessing the most viable method for automation. The significance of this research will work to develop a practical and accessible *in situ* methodology using automated cameras to assess Arctic climate change through observational science. Ultimately, the results from this study will provide critical insight into the spatial conditions of lake ice and snow in the study region under the current climate regime, which will support better projections under warming climate conditions in the Arctic.

[1] Department of Geography, Geomatics and Environment, University of Toronto Mississauga, Mississauga, ON, Canada

Projection of Lake Ice Thickness and Phenology under Representative Concentration Pathways (RCP) Scenarios: Great Slave and Great Bear Lakes, Northwest Territories

ARIANA MANSINGH[1], HOMA KHEYROLLAH POUR[1], AND GIFTY ATTIAH[1]

ABSTRACT

Cold region lakes are key ecosystems acting as sentinels of climate change. The increased severity of climate change is causing a demand for enhanced environmental understanding. Thus, knowledge gaps within the field must be addressed. Currently, the understanding of far future implications of climate change on lake ice thickness and phenology is limited and is challenging in projecting a spatially and temporally extensive lake ice thickness and phenology. This project is taking place on Great Bear Lake and Great Slave Lake, Northwest Territories, to better understand the long-term implications of greenhouse gas emissions and their influence on the future lake ice formation and duration, using a spatially distributed lake ice model, CLIMoGrid. To do so, Representative Concentration Pathways (RCPs) are used; these portray different scenarios of greenhouse gas concentrations and emissions. Data from a pre-existing model called MRI-CCGM is used based on the different RCPs to force CLIMoGrid, such as cloud cover, air temperature, humidity, snowfall, and wind speed. The model was able to project future ice thickness and phenology based on the future RCPs scenarios with more rapid changes under the RCP8.5 scenario by 2085. The overall outcome of this research was to produce a trend analysis of ice thickness and phenology and project future changes that consider different RCPs and climate change's long-term implications. This research provides support to the emerging spectrum of climate change issues throughout cold regions.

[1] Department of Geography and Environmental Studies, Wilfred Laurier University, Waterloo, ON, Canada

Comparaison de la Performance de Modèles Empiriques de Bilan de Masse Glaciologique en Réponse au Changement Climatique

LISA MICHAUD[1] AND CHRISTOPHE KINNARD[1]

ABSTRACT

La prévision de l'évolution du bilan de masse glaciologique se fait à l'aide de modèles qui reproduisent, à l'aide de données météorologiques, les mesures prises sur le terrain. Les modèles physiques sont efficaces pour faire des prévisions à long terme, mais nécessitent de nombreuses données météorologiques, pas toujours disponibles. Les modèles empiriques, qui nécessitent peu de données météorologiques, sont donc souvent utilisés. Les études comparant la performance des modèles empiriques concluent que ceux qui représentent mieux le bilan de masse des dernières décennies auront ègalement les meilleures projections durant le prochain siècle. Toutefois, les conditions dans lesquelles ces modèles ont été calibrés ne seront plus les mêmes dans les prochaines décennies en raison du changement climatique. Les conclusions de ces études peuvent être erronées puisque les modèles empiriques sont moins performants lorsqu'ils sont utilisés hors de leurs conditions de calibration. Afin d'explorer cette problématique, cette étude compare les résultats de modèles empiriques directement dans les conditions climatiques des prochaines décennies. Les modèles empiriques ont été calibrés en comparant leurs résultats avec ceux du modèle physique de Hock et Holmgren (2005) modifié par Kinnard *et al.* (2022) pour le glacier Saskatchewan, en Alberta. Les projections du modèle physique sont considérées comme étant les 'valeurs réelles' de bilan de masse des prochaines décennies. Trois méthodes de calibration sont testées afin de voir l'impact du choix de la méthode sur les projections des modèles empiriques. Cet impact est peu mentioné dans les études sur le sujet. Pourtant, cela affecte les valeurs optimales des paramètres des modèles lors de la calibration, ce qui a un impact sur les valeurs de bilan de masse prédites. Nous avons également constaté que la performance des modèles empiriques est plus affectée par les changements de température que par les changements de taux de précipitations dans différents scénarios climatiques.

[1] Département des science de l'environnement, Université du Québec à Trois-Rivières, Trois-Rivières, QC, Canada

Forward Modeling of SAR Backscatter during Lake Ice Melt Conditions using the Snow Microwave Transfer (SMRT) Model

JUSTIN MURFITT[1], CLAUDE DUGUAY[1,2], JUHA LEMMETYINEN[3], AND GHISLAIN PICARD[4]

ABSTRACT

Lake ice is a key component of northern landscapes. It plays a critical role in local energy balances and provides crucial socioeconomic services such as travel between communities and transportation of goods during winter months. However, changing climate is impacting the duration of ice seasons and the thickness of ice covers. Additionally, increasing temperatures lead to an increasing number of melt events throughout the season, resulting in the formation of more snow ice and slush layers. The presence of slush layers can pose challenges for mapping lake ice using active microwave data (i.e., synthetic aperture radar) and be a risk to those who use ice cover as an essential travel route. Using radiative transfer modeling we can better understand the connection between these events, lake ice properties, and remote sensing; however, exploration has been limited. This study will focus on Lake Oulujõrvi during the 2020-2021 ice season and use detailed field data collected over three dates. Field data collected in late January and early March are representative of dry conditions, while field data collected in late March shows evidence of increased water content. Snow density, snow depth, microstructure data, and ice thickness data collected during the field campaign will be used to parameterize the Snow Microwave Radiative Transfer (SMRT) model. SMRT will be used to conduct forward modeling simulations of the ice cover during these dates. The focus of these simulations is to understand how changes in snow and ice properties during melt events impact backscatter and how different conditions impact the dominant scattering interface (i.e., ice-water, snow-ice, and ice-air). The results of this modeling will serve as an important basis for further improving the parameterization of radiative transfer models throughout the ice season.

[1] Department of Geography and Environmental Management, University of Waterloo, Waterloo, ON, Canada
[2] H20 Geomatics, Kitchener, ON, Canada
[3] Finnish Meteorological Institute, Helsinki, Finland
[4] Institut des Géosciences de l'Environnement, Université Grenoble Alpes, Grenoble, France

Heat Wave Impacts on Glacier Mass Balance, Glacier Runoff, and Salmon in Nooksack River, Washington

MAURI PELTO[1], MARIAMA DRYAK[2], AND JILL PELTO[3]

ABSTRACT

In the North Cascade Range, Washington, USA, glacier runoff is a major source of streamflow during the summer low-flow season and mitigates both low-flow and high-water temperatures, particularly during summer heat waves. A thirty-eight-year record (1984-2021) of glacier mass balance and areal extent measurement indicate a significant glacier response to climate change in the North Cascades, Washington, that has led to declining glacier runoff. The ameliorating role of glacier runoff on discharge and water temperature is examined during 23 late-summer heat wave events from 2010-2021 including the early summer 2021 Pacific Northwest heat wave. The heat waves are characterized by local temperature data. Glacier runoff is determined using synchronous observations of glacier ablation on Sholes Glacier and stream discharge immediately below Sholes Glacier. The observed ablation rate is applied to glaciers across the North Fork Nooksack watershed, providing daily glacier runoff discharge to the North Fork Nooksack River. This is compared to observed daily discharge and temperature data of the North Fork Nooksack River and the unglaciated South Fork Nooksack River from the USGS. During the 23 warm weather events the discharge increased an average of +23% in the North Fork and decreased an average of 20% in the South Fork. For water temperature the mean increase was 0.7 °C in the North Fork and 2 °C in the South Fork. For the North Fork glacier runoff production was equivalent to 34% of the total discharge during the 23 events. Heat waves that occurred after significant bare firn and ice was exposed on the glacier surface led to peak glacier discharges. Ongoing climate change will cause further decreases in summer baseflow and an increase in water temperature potentially exceeding tolerance levels of several Pacific salmonid species.

[1] Nichols College, Dudley, MA, USA
[2] CIRES, University of Colorado Boulder, Boulder, CO, USA
[3] Portland, ME, USA

Evaluation of Snow Depth Derived from Ground Penetrating Radar on Canadian Subarctic Lakes

ALICIA POUW[1], HOMA KHEYROLLAH POUR[1], AND ALEX MACLEAN[1]

ABSTRACT

In subarctic regions, a realistic representation of the snow depth and distribution on lake ice is important for climate change studies and lake ice modeling. Small changes in the surface-atmosphere energy balance could alter snowpack dynamics and can have a profound impact on lake ice thickness and formation. Snow depth observations on lake ice are sparse and mostly restricted to point measurements. Therefore, developing accurate methods of estimating the snow depth and spatial distribution across lake ice is important, but challenging. In this study, we applied an algorithm to estimate snow depths on lake ice using ground penetrating radar (GPR) two-way travel-times (TWT) over four small subarctic freshwater lakes (Landing Lake, Long Lake, Finger Lake, Vee Lake), located north of Yellowknife, Northwest Territories. High spatial resolution (~9 cm) observations along transects totaling ~38km were acquired using a 1000 MHz sensor. We derived the snow depth by automatically picking the GPR TWT of the snow-ice interface and combining it with a measured average snow density for each lake. The accuracy of the derived snow depth is assessed using a snow depth GPS magnaprobe. Limited by the spatial accuracy of the snow depth magnaprobe and GPR, a 6 m radius was applied around each *in situ* snow depth measurement, where it is confirmed the GPR estimated the snow depth with an $R^2 = 0.62$, mean bias error of -0.33 cm, and root mean square error of 1.72 cm, on average for all four lakes. The results showed that this algorithm can improve the accuracy of snow depth and distribution retrieval data on lake ice which is essential for climate change studies and the lake ice modeling community.

[1] Department of Geography and Environmental Studies, Wilfred Laurier University, Waterloo, ON, Canada

Improving Lake Ice Simulations in Canada based on Lake Size

ALEXIS ROBINSON[1] AND LAURA C. BROWN[1]

ABSTRACT

Lake coverage in Canada is estimated to be between 15-40% depending on location and many experience some form of ice cover throughout the year. Research shows that lake ice duration is shortening in response to a warming climate. Lake ice models are an important tool that can be used to study spatial and temporal changes to lake ice regimes under a changing climate. The Canadian Lake Ice Model (CLIMo) is a well-tested one-dimensional thermodynamic freshwater ice model that has been used to successfully simulate Arctic and sub-Arctic lake ice cover. Research from central Ontario, Canada, shows that adjusting the albedo parameterization in CLIMo (CLIMo-Temperate) results in a better representation of temperate region lake ice cover. This study uses CLIMo and CLIMo-Temperate to simulate lake ice-on and ice-off dates for 174 lakes across Canada that have historical ice records available. Model inputs include meteorological data obtained from Environment and Climate Change Canada with simulations validated using the historical ice records in the Canadian Ice Database and the Global Lake and River Ice Phenology Database. Results for small (< 1 km^2) have a mean absolute error (MAE) of 2 to 9 days. For medium (1-100 km^2) and large sized lakes (> 100 km^2) the MAE ranges from 3 to 37 days for ice-off. The results show a satisfactory agreement between field measurements and modeled dates for lakes < 1 km^2, however, there is substantial variation in ice-off dates for lakes > 1 km^2. The findings show the importance of accounting for lake size and provide a basis to derive a lake size adjustment factor. This size factor will improve the accuracy of one-dimensional lake ice models, which will lead to improvements for global and regional climate models, numerical weather forecasting, and future climate projections.

[1] Department of Geography, Geomatics and Environment, University of Toronto Mississauga, Missasauga, ON, Canada

Recommendations to Enhance Hydrological Models for Improved Estimates of Climate Impacts on Northern Waters

ROBIN THORNE[1], PHILIP MARSH[1], DAVID RUDOLPH[2], CHRISTOPHER SPENCE[3], OLIVER SONNENTAG[4], JEFFREY MCKENZIE[5], AND AARON BERG[6]

ABSTRACT

The rapidly changing Arctic climate is impacting the interactions between surface water, supra-permafrost water, snow, vegetation, lakes, and streamflow, particularly as ongoing climate warming continues to thaw the permafrost. Across the Northern Water Futures (NWF) study domain, which is a Northwest Territories-focused consortium, spatial and temporal variability in active layer thickness, the occurrence of taliks, soil infiltration, subsidence, and the contribution of groundwater flow to streamflow, are all expected to be enhanced. A key challenge for Arctic models is dealing with freeze/thaw conditions and the high spatial variability in terrain and vegetation. Hydrologic models are being developed in NWF and other research groups around the globe, but there are many difficulties and approaches in developing such models.

Integrated surface and subsurface hydrological models are critical tools to investigate the impacts of warming on Arctic hydrology over the coming decades and centuries. First, we will review a suite of available hydrological models for this purpose, with a focus on interactions between permafrost, groundwater, and streamflow. We will consider potential limitations due to the lack of sophisticated 3D permafrost/groundwater model components and limitations in modeling at the high spatial resolution required. Second, we will use a case example in changes to the development of Arctic models using examples from the Trail Valley Creek (TVC) research watershed, north of Inuvik, NWT. Finally, we will make recommendations to enhance model development as needed, such as the inclusion of components from cryohydrogeological and lake process models, to address changes in Arctic hydrology at the scales required to address societal needs.

---

[1] Department of Geography and Environmental Studies, Wilfred Laurier University, Waterloo, ON, Canada

[2] Department of Earth and Environmental Sciences, University of Waterloo, Waterloo, ON, Canada

[3] Environment and Climate Change Canada, Saskatoon, SK, Canada

[4] Département de géographie, Université de Montréal, Montréal, QC, Canada

[5] Department of Earth and Planetary Sciences, McGill University, Montréal, ON, Canada

[6] Department of Geography, Environment and Geomatics, University of Guelph, Guelph, ON, Canada

Variability in Thermokarst Lake Size, Elevation, and Connectivity in the Western Canadian Arctic

ROSAMOND TUTTON[1], PHILIP MARSH[1], AARON BERG[2], RODERICK MELNIK[3], BEN DEVRIES[2], AND BRANDEN WALKER[1]

ABSTRACT

Large parts of the Arctic are densely covered by vast numbers of thermokarst lakes, ranging in size and stream network connectivity. Such stream-lake networks are poorly studied in the Canadian Arctic. However, a warming climate coupled with permafrost thaw and precipitation change will significantly influence the hydrology of these lakes, including changes in water surface elevation (WSE) and area. Understanding the impact of controls such as climate, snow-cover, vegetation, permafrost thaw, beavers, and hydrological connectivity on WSE requires a combination of field and remote sensing observation and integrated snow, permafrost, runoff, and lake hydrologic models. This paper will present the first step in putting such an observing system in place. We will report on observations of lake WSE across 50 lakes north of Inuvik, NWT. These lakes are located in two contrasting watersheds - Trail Valley Creek (TVC) and Hans Creek (HC). TVC is lake poor while HC is lake rich with a dense network of connected streams and lakes. In addition, HC is dominated by large networks of ice-rich polygonal terrain that is prone to thawing with dramatic impacts on these lakes. These thermokarst lakes are important habitats for migratory birds, fish, and mammals and vital to sustaining hunting, fishing, trapping, transportation, freshwater resources, and recreation of those that live in the north. This project will contribute to a Canadian Space Agency project aimed at validation of WSE estimates from the Surface Water and Ocean Topography (SWOT) mission to be launched later in 2022 and development of integrated lake-stream network models.

[1] Department of Geography and Environmental Studies, Wilfred Laurier University, Waterloo, ON, Canada
[2] Department of Geography, Environment and Geomatics, University of Guelph, Guelph, ON, Canada
[3] Department of Mathematics, Wilfred Laurier University, Waterloo, ON, Canada

Why do Simulated Trends of Arctic Sea Ice Drift Speed go from Positive in the 20th Century to Negative in the 21st Century?

JAMIE WARD[1] AND NEIL TANDON[1]

ABSTRACT

In addition to its importance for ecosystems and shipping activity, motion of Arctic sea ice might also strongly influence the rate of Arctic sea ice extent decline. Sea ice motion is driven by a combination of wind stress, ocean stress, sea ice internal stress, and sea surface height (SSH) gradients. Buoy observations indicate that sea ice drift speed has been increasing since the 1980s due to increased sea ice deformation associated with reduced sea ice thickness. Modern coupled climate models generally produce similar behaviour during this time period, but puzzlingly, most climate models also project that drift speed will decline over the middle and late 21st century. In this study, we investigate the reasons for this sign change in simulated Arctic drift speed trends using an ensemble of 17 models participating in the Coupled Model Intercomparison Project phase 6 (CMIP6) under the shared socioeconomic pathway (SSP) 585 warming scenario.

We find that, with the exception of one model (NorESM2-MM), all models in our ensemble project decreasing Arctic-average September sea ice drift speeds in the mid-to-late 21st century. Our analysis indicates that these declining drift speeds are primarily due to reduced SSH gradients over the Arctic Ocean. Trends in sea ice internal stress have also been examined, and these do not explain the decline in sea ice drift speed. Analysis of CMIP5 in an earlier study also indicates that changes in wind stress and ocean stress likely do not explain these trends. Altogether, our results suggest that, in addition to understanding sea ice internal stresses, gradients of sea surface height must also be considered in order to accurately project long-term changes in sea ice motion.

[1] Department of Earth and Space Science and Engineering, York University, Toronto, ON, Canada

Multi-Method Approach to Inventorying Rock Glaciers and Features of Interest in Banff and Jasper National Parks, Alberta, Canada

MISHELLE WEHBE[1], PHILIP BONNAVENTURE[1], AND ROBERT WAY[2]

ABSTRACT

Rock glaciers are perennially frozen masses of ice and unconsolidated material that creep downslope resultant from internal, cyclic freeze-thaw mechanisms and weight due to gravity. These features are often tongue-shaped, lobate landforms, with lateral and frontal margins, that often contain longitudinal or transverse flow features within their surface consisting of poorly sorted, angular rock debris. Although rock glaciers are abundant geomorphological features in the alpine periglacial environments of the Rocky Mountains, their spatial distribution and characteristics are largely unknown. As rock glaciers contain frozen fresh water and can be potential geohazards, inventories are crucial in the assessment of the activity status and distribution of these landforms. This inventory will also provide an estimate on the potential cubic meters of water equivalent storage within the rock glaciers for cryospheric reserves as freshwater depletes. To date, over 800 intact (active/inactive) rock glaciers were successfully identified within the study areas, as well as over 204 features of interest requiring further validation of surface kinematics and morphometric quantification. Grid-based manual inventorying of these features was completed using high-resolution satellite imagery that is readily available through the ESRI World Imagery Base Layer, then subsequently verified manually with Google Earth Pro. This proved crucial and supports the idea of both multi-temporal and multi-method approaches to the inventorying of rock glaciers and features of interest within the alpine terrains of Canada to ensure high accuracy in inventorying and for long-term monitoring feasibility. This work represents the first component of our rock glacier monitoring network within Canada, as there currently are none to date.

[1] Department of Geography and Environment, University of Lethbridge, Lethbridge, AB, Canada
[2] Department of Geography and Planning, Queen's University, Kingston, ON, Canada

Session #4: Novel Approaches in Seasonal Snow: Field Measurements, Remote Sensing, Modeling, and Data Assimilation and Analytics

Retrieval of Snow Water Equivalent from SWESARR Measurements in Grand Mesa Colorado, SnowEx 2020

MICHAEL DURAND[1] AND BATUHAN OSMANOGLU[2]

ABSTRACT

One critical question in applying radar measurements to estimate SWE is to assess how much prior information is needed to accurately perform the retrieval. We apply retrieval algorithms to SWESARR measurements of radar backscatter made in February 2020, as part of the SnowEx campaign. The algorithm removes background backscatter from a snow-off measurement in November. The algorithm uses a physically-based microwave radiative transfer model to relate the snow and backscatter properties, and iteratively computes the optimal estimates. Both one- and two-layer configurations are explored. The algorithm is Bayesian in that it uses *a priori* information on SWE, as well as on the microstructure correlation length. We explore three configurations: one in which no *a priori* information is used, one in which generic prior information is used, and a third where local prior information specific to Grand Mesa is used. We hypothesize that the two-layer model with local priors performs best, but that the two-layer model with generic priors still performs adequately. Demonstration that generic priors lead to adequate performance would strengthen the case that global SWE can be estimated from satellite measurements of Ku and X band radar.

[1] School of Earth Sciences, The Ohio State University, Columbus, OH, USA
[2] NASA Goddard Space Flight Center, Greenbelt, MD, USA

Predicting Surface Density using Snow Models and Assimilation for Wildlife Applications

MICHAEL DURAND[1], JACK DECHOW[1], JESSICA LUNDQUIST[2], LAURA PRUGH[3], BEN SULLENDER[3], CASSIE LUMBRAZO[2], KATIE BREEN[3], AND CALUM CUNNINGHAM[3]

ABSTRACT

Surface layer snow density is correlated with the depth that animals sink into snow, which in turn controls their mobility and ultimately impacts predator-prey dynamics. Snow model density parameterizations have a long history, and thus are a promising tool to predict wildlife dynamics. However, surface snow density is challenging to predict in areas that experience melt-refreeze metamorphism in part because it is challenging to accurately predict brief periods of snowmelt. This suggests the use of spaceborne land surface temperature (LST) measurements to correct the model. Here, we explore ability of the NoahMP snow model in conjunction with the NASA Land Informatic System Framework (LIS or LISF) in order to predict surface snow density with and without LST assimilation at a range of spatial resolutions from 1 km to 30 m in Washington State. We evaluate against field surveys of surface snow density made in winter 2021 and 2022. Without assimilating LST, we have successfully predicted whether the density was greater or less than 275 kg m^3 for 80% of snow pits. We have now begun to assimilate MODIS LST measurements in order to correct the model predictions of meltwater within the snowpack and improve estimates. Furthermore, we have begun to analyze the SWE estimates from the model as well. We use the improved density results of the model and assimilation study to assess the possible utility of snow modeled density to predict snow wildlife dynamics.

[1] School of Earth Sciences, The Ohio State University, Columbus, OH, USA
[2] Department of Civil and Environmental Engineering, University of Washington, Seattle, WA, USA
[3] School of Environmental and Forest Sciences, University of Washington, Seattle, WA, USA

Quantifying Volumetric Scattering Bias in ICESat-2 Altimetry over Snow-Covered Surfaces

ZACHARY FAIR[1], MARK FLANNER[2], CARRIE VUYOVICH[1], TOM NEUMANN[1], BENJAMIN SMITH[3], AND ADAM SCHNEIDER[4]

ABSTRACT

The ICESat-2 mission has collected global measurements of surface elevation for over three years. The sole onboard instrument, the Advanced Topographic Laser Altimeter System (ATLAS), operates at 532 nm, and ice and snow absorb weakly at this wavelength. Previous modeling studies found that aged or melting snow may induce 0.45 m in altimetry bias, though there has yet to be a formal assessment on ICESat-2 acquisitions during the melting season. There is a recent interest in using ICESat-2 data for snow applications, so it is vital to quantify these biases to ensure accurate measurements over snow-covered surfaces.

We perform two case studies that utilize lidar altimetry data from ICESat-2 and the Airborne Topographic Mapper (ATM) over the Greenland Ice Sheet to quantify volumetric scattering bias in snow. The case studies are conducted near the end of the melting season in September 2019. A Monte Carlo photon-tracking model and snow grain sizes derived from ATM waveforms and the Next Generation Airborne Visible/Infrared Imaging Spectrometer (AVIRIS-NG) are used to attribute ICESat-2 biases to snow.

Our results indicate that ICESat-2 and ATM experience centimeter-level bias across the Greenland Ice Sheet, and we generally see increases in bias with increasing grain size. Relative to the ATM 1064 nm beam, we find that mean biases derived from ICESat-2 and ATM over the Greenland Ice Sheet are 6.9 cm and 6.01 cm, respectively. These biases agree best with modeled results when simulated snow density is 250-350 kg m^{-3}. Although the ICESat-2 errors are within mission accuracy requirements, we cannot rule out more significant errors in regions of melting snow, particularly over forested or sloped terrain. We expect to continue this study over mid-latitude field sites in support of the SnowEx mission, with a bias correction algorithm and regional snow depth retrievals as future research goals.

[1] NASA Goddard Space Flight Center, Greenbelt, MD, USA

[2] Department of Climate and Space Sciences and Engineering, University of Michigan, Ann Arbor, MI, USA

[3] Polar Science Center, University of Washington, Seattle, WA, USA

4 Department of Earth System Science, University of California, Irvine, Irvine, CA, USA

Using Machine Learning to Estimate Snow Cover from Ground Surface Temperature Measurements

ANIKA FORGET[1], ROBERT WAY[1], AND ROSAMOND TUTTON[2]

ABSTRACT

In Arctic and Subarctic landscapes, snow cover can significantly affect ground temperatures and can lead to changes in permafrost distribution. Snow cover and snow thickness are highly heterogeneous through time and space leading to profound local scale variability and difficulties in representing snow across large areas. Owing to the high cost of snow monitoring equipment, many researchers working in the north use ground surface temperature (GST) loggers distributed across an area to estimate snow characteristics, including snow onset and melt dates. However, existing methods for evaluating snow cover from GST data depend on arbitrary thresholds selected by trial and error. Such thresholds can vary widely between sites due to snow cover variability, as well as within sites depending on vegetation cover and soil composition. This study established a novel method using supervised machine learning (random forest) to predict daily snow cover from GST data, removing the need for local threshold selection. This random forest method was built using simulations derived from the Northern Ecosystem Soil Temperature model that were calibrated at the Subarctic Pinware River Hills (PRH) research basin in coastal Labrador. Estimates of snow cover derived via application of the machine learning algorithm to GST data were compared against snow thickness observations at PRH from multiple co-located sources. Our results indicate that this novel method more accurately predicts total snow cover days, snow onset, and melt dates compared to previously used threshold-based approaches. These early results show tremendous potential for reducing error in deriving snow cover information from GST observations and can enable more reliable snow cover estimation across larger areas and differing ecotypes. Insights gained from this research could be applied to the large networks of GST observations currently being assembled for the northern circumpolar regions to improve snow cover modeling.

[1] Department of Geography and Planning, Queen's University, Kingston, ON, Canada
[2] Department of Geography and Environmental Studies, Wilfred Laurier University, Waterloo, ON, Canada

The Airborne Cyrosphere-Observing Synthetic Aperture Radar System (CryoSAR): A Snow, Soil, Sea Ice, and Lake Ice Observing System

RICHARD KELLY[1], AARON THOMPSON[1], ADRIANO META[2], JIM HODGSON[3], ALEXANDRE LANGLOIS[4], ALEXANDRE ROY[5], LAURA BROWN[6], AARON BERG[7], AND CHRIS DERKSEN[8]

ABSTRACT

The airborne Cryosphere-Observing SAR (CryoSAR) system is a Ku- and L-band polarimetric synthetic aperture radar (SAR) system designed to conduct observations of snow and ice on land and over water bodies. The CryoSAR system is designed as a fully polarimetric SAR with the capability to conduct single or repeat pass observations for interferometric SAR (InSAR) applications. There is significant interest in the Ku- (13.5 GHz) and L-band (1.3 GHz) InSAR measurements of snow on land and water as tools to estimate snow water equivalent, a key variable in water resource management applications and in climate change studies. The CryoSAR radars can be operated independently or together. They can also be deployed on a relatively small aircraft, such as a Cessna 208, which is generally available across North America, Europe, and beyond, making the system relatively agile in its deployment. An adjustable mounting system has been designed to enable the instrument to be installed on a 208 platform from inside and at specified look angles. In winter 2022, the first deployment of the system was planned in Ontario as part of a Canadian Space Agency-funded project. Flight paths were identified over the Haliburton Highlands and in Powassan. While Federal certification could not be completed in time, and the aircraft could not be deployed for flight overpasses, this presentation provides a description of the radar operation and what can be achieved. It also describes a snow field campaign that was conducted to support potential aircraft acquisitions. A combination of traditional field observations of snow properties, and detailed state-of-the-art measurements of microstructure properties were made to quantify the snowpack bulk and stratigraphic characteristics of the snow at the different field sites. Once federally-certified, the CryoSAR will be deployed in Ontario and Alberta in the 2022-23 winter season.

[1] Department of Geography and Environmental Management, University of Waterloo, Waterloo, ON, Canada

[2] Metasensing, Milano, Italy

[3] Lake Central Air Services, Gravenhurst, ON, Canada

[4] Département de géomatique appliquée, Université de Sherbrooke, Sherbrooke, QC, Canada

[5] Département des Sciences de l'Environnement, Université du Québec à Trois-Rivières, Trois-Rivières, QC, Canada

[6] Department of Geography, Geomatics and Environment, University of Toronto Mississauga, Mississauga, ON, Canada

[7] Department of Geography, Environment and Geomatics, University of Guelph, Guelph, ON, Canada

[8] Environment and Climate Change Canada, Toronto, ON, Canada

Advances in X- and Ku- Band Radar Algorithms for SWE Retrieval by Future Satellite Missions

EDWARD KIM[1], DK KANG[1,2], FIROZ BORAH[3], AND LEUNG TSANG[3]

ABSTRACT

A leading measurement technique for a future SWE satellite mission is radar volume scattering at X and Ku bands (10 and 17 GHz). For example, the European CoReH2O concept was based around this, but unfortunately was not selected. Many key advances have been made since then, both in the theoretical domain as well as in the area of validation against field observations - for example, from SnowEx. Using those advances, there is a NASA mission concept being developed around this technique.

At the core of the current algorithm, and one of the key advances, is a dense media radiative transfer (DMRT) model in a simplified parameterized form. Other advances include subtraction of background scattering and partitioning by the scattering albedo. The retrieval solution is determined through a cost function minimization approach after accounting for surface scattering from the snow-soil interface, low vegetation, and the scattering albedo.

We will show validation results using both tower and airborne data and discuss the options for obtaining the necessary input parameters.

[1] NASA Goddard Space Flight Center, Greenbelt, MD, USA
[2] ESSIC, University of Maryland, College Park, MD, USA
[3] Department of Electrical Engineering and Computer Science, University of Michigan, Ann Arbor, MI, USA

Spatiotemporal Assessment of Snow Density and Snow Water Equivalent (SWE) using the Cone Penetration Test (CPT) and/or Ground Penetrating Radar (GPR)

ADRIAN MCCALLUM[1]

ABSTRACT

In situ tests such as the Cone Penetration Test (CPT) are useful for establishing stratigraphy and physical properties of frozen geomaterials such as snow and ice. Particularly, snow density can be estimated using CPT and these data can then be spatially extrapolated using Ground Penetrating Radar (GPR). Further, GPR can potentially be used in isolation to assess spatiotemporal variability of snow surface and sub-surface snow density, enabling changes in snow water equivalent (SWE) to be estimated. Almost 100 CPTs conducted in Antarctica demonstrated the potential to assess snow density and mass balance variability, directly from CPT data, and GPR testing in the Australian Alps suggested the ability to estimate both surface and sub-surface snow-density, by assessing amplitude variation from air-coupled GPR surveys. Combination of these tests may enable efficient spatiotemporal estimation of snowpack density and SWE variability.

[1] School of Science, Technology and Engineering, University of the Sunshine Coast, Slippy Downs, QLD, Australia

Leveraging Adaptive Viewing to Improve the Efficacy of Space-Borne Satellite Retrievals of Terrestrial Snow

COLIN MCLAUGHLIN[1], BARTON FORMAN[1], AND LIZHAO WANG[1]

ABSTRACT

Space-borne sensors for snow estimation traditionally employ a "fixed" viewing strategy that applies a static look angle between the sensor and the observed portion of Earth's surface. However, a fixed viewing strategy often misses the target of interest (e.g., views snow-free valley rather than snow-capped peak). An alternative to fixed viewing is "adaptive" viewing such that the satellite-based sensor can slew in order to preferentially view the target of interest. This project developed an algorithm to model adaptive viewing strategies in order to explore how adaptive viewing by a sensor can maximize the information content of satellite observations. The synthetic retrievals developed in this project are ultimately used in an observing system simulation experiment (OSSE) to quantify how, and to what degree, adaptive sensor viewing has the potential to improve the characterization of terrestrial snow mass.

In this project, the adaptive viewing model is used to generate synthetic retrievals of snow depth and SWE. The adaptive viewing strategy is highly flexible but remains constrained to user-specified parameters such as satellite trajectory, swath width, maximum rate of slew angle, and maximum extent of slew angle. Results from the adaptive viewing strategy are compared relative to those from the fixed viewing strategy. When using a hypothetical LiDAR sensor with a wide swath width, for example, adaptive viewing model captures 30-50% more of the target of interest relative to the fixed viewing strategy. This investigative approach is repeated over a variety of satellite trajectories, swath widths, and slewing rates.

[1] Department of Civil and Environmental Engineering, University of Maryland, College Park, MD, USA

Assimilation of GRACE/GRACE-FO Terrestrial Water Storage Retrievals to Improve Snow Mass Estimates across North America

ALIREZA MOGHADDASI[1] AND BARTON FORMAN[1]

ABSTRACT

Accurate estimation of snow mass, and its change in space and time, is necessary for the optimal management of freshwater resources, especially in regions of complex terrain. Assimilation of remotely-sensed observations into a land surface model (LSM) can help better characterize snow mass. Gravimetric observations collected by the Gravity Recovery and Climate Experiment (GRACE / GRACE-FO) are a useful complement to more traditional means of snow remote sensing (i.e., passive microwave radiometry) because, unlike radiometers, gravimetry does not suffer from signal saturation in the presence of deep and/or wet snow. However, gravimetry has its own limitations, primarily in the form of coarse spatial and temporal resolutions.

The Noah-MP land surface model without the benefit of assimilated observations (a.k.a., Open Loop; OL) is first investigated within the NASA Land Information System (LIS) to estimate hydrologic states and fluxes (including snow) across North America. GRACE-based retrievals of terrestrial water storage (TWS) are then assimilated into the Noah-MP model using an ensemble Kalman filter in order to dynamically update model-derived estimates of snow water equivalent (SWE) based on the TWS retrievals. To evaluate changes in modeled snow with and without assimilation, model results are validated against ground-based measurements of snow mass across North America obtained from the Canadian Snow Water Equivalent dataset (CanSWE), the United States SNOTEL network, and the European GlobSnow product. Preliminary results suggest improvement in terms of both snow amount and timing of snow accumulation and ablation in regions where snow is a significant contributor to the hydrologic cycle. Findings from this study will help water resources managers by providing useful information that ultimately leads to better characterization of snow in regions with limited ground-based stations and in places where snow is variable across time and space.

[1] Department of Civil and Environmental Engineering, University of Maryland, College Park, MD, USA

Climate Driven Changes in Snowpack: Simulations (1970s to 2020) for the Bay of Quinte, Ontario, Canada

AGNES RICHARDS[1], FÉLIX OUELLET[1], ÉRIKA BOISVERT-VIGNEAULT[2], XIAOYUAN GENG[3], AND ALEXANDRE LANGLOIS[2]

ABSTRACT

SNOWPACK simulations that capture climate driven changes in snow water equivalent (SWE) and snow cover will be presented from the 1970s to 2020 for the entire Bay of Quinte watershed located in Ontario, Canada. This is an Area of Concern, listed under the Great Lakes Water Quality Agreement due to consistent harmful algal blooms resulting from excessive phosphorus runoff from the watershed. Future management actions should consider the effects of climate change, such as increased frequency and intensity of storm events. These events will amplify erosion and leaching of phosphorus from soils, while ruptured vegetation from freeze-thaw will also contribute to further leaching. Our simulations focus on the changes in snow parameters, specifically SWE, which drive snow melt. Other parameters such as snow density will also be discussed. Preliminary simulations show a substantial change in SWE, especially after 2000. The implications of changes in SNOWPACK on runoff and nutrient loadings will be discussed.

[1] Environment and Climate Change Canada, Burlington, ON, Canada
[2] Département de géomatique appliquée, Université de Sherbrooke, Sherbrooke, QC, Canada
[3] Agriculture and Agri-Food Canada, Ottawa, ON, Canada

Evaluating Passive Microwave Snowmelt Detection Methods with Ground Snow Observations

ANGELA RIENZO[1] AND SAM TUTTLE[1]

ABSTRACT

Snowmelt is an important hydrological factor in many areas in North America, especially in regions that rely on seasonal snowmelt to replenish water supply. Rapid spring melt can also lead to destructive river flooding. Because snowpack characteristics can vary significantly in different regions and climates, accurate global snowmelt modeling may require assimilation of both remote sensing and ground observations. The purpose of this investigation is to determine the level of insight that passive microwave remote sensing techniques can provide into complex snowmelt processes. We compare snowmelt events detected from passive microwave brightness temperatures (Tuttle & Jacobs, 2019, Water Resources Research) to daily observed snow water equivalent (SWE) and snow depth data from over 600 SNOTEL stations to determine how much information satellite data can provide about snowmelt processes. Specifically, we evaluate whether the satellite-detected melt events align with SWE decreases (water loss from the snowpack), or snow depth decreases without decreases in SWE (possible midseason melt events in which the melt is refrozen in the snowpack). Comparisons from regional subsets of SNOTEL stations are further analyzed to determine if the satellite snowmelt detection method has a higher accuracy in certain snowpack conditions and climates.

[1] Department of Earth and Environment Sciences, Syracuse University, Syracuse, NY, USA

An Accurate Global Daily Snow Cover and Albedo Product from MODIS and VIIRS

KARL RITTGER[1], KEITH MUSSELMAN[1], MCKENZIE SKILES[2], NED BAIR[3], MARY J. BRODZIK[4], MARK SERREZE[5], STEPHANIE ABEGG[6], AND WILLIAM KLEIBER[6]

ABSTRACT

Water managers need accurate observations of snow cover and albedo to make decisions for a diverse set of applications. Remotely sensed snow cover and albedo products that are currently available do not meet operational requirements for several reasons. We have partnered with snow remote sensing end users who serve diverse needs of national and international water resource decision makers. With their guidance, we create and provide daily gap-filled snow cover and albedo, including impacts of light absorbing particles. The products account for off-nadir views, snow under the forest canopy, and use cloud filtering techniques not employed in existing products. Using algorithms shown to perform consistently across sensors—specifically MODIS and VIIRS—we will process the historical daily record and produce data in near real-time with a sub-daily latency period. This project will complete the transition of the data processing, archiving, and distribution to the National Snow and Ice Data Center. The transition will ensure the continued production of snow cover and snow albedo products for the lifetime of these sensors.

While expanding these products to global coverage, we recognize that snow surface properties vary at a much finer scale than the resolution of MODIS and VIIRS. While the combination of Landsat 8 & 9 and Sentinel 2a&b approach near-daily resolution, they will need to operate for several years together to create a long-term record. To create a daily Landsat resolution dataset for previous periods we fuse the snow and albedo products using a random forest model trained on MODIS, Landsat 8, and a suite of geophysical predictor variables. We develop separate models to address the issue of missing data within the MODIS product caused by its coarse spatial resolution, allowing for more complete spatial fusion coverage, especially near the snow line where snowmelt is prominent.

[1] Institute of Arctic and Alpine Research, University of Colorado Boulder, Boulder, CO, USA
[2] Department of Geography, University of Utah, Salt Lake City, UT, USA
[3] Earth Research Institute, University of California, Santa Barbara, Santa Barbara, CA, USA
[4] National Snow and Ice Data Center, Boulder, CO, USA
[5] Department of Geography, University of Colorado Boulder, Boulder, CO, USA
[6] Department of Applied Mathematics, University of Colorado Boulder, Boulder, CO, USA

DLR Global SnowPack – Possible Applications of the Near Real-Time Product

SEBASTIAN RÖßLER[1] AND ANDREAS DIETZ[1]

ABSTRACT

For over 20 years, the MODIS sensors on Terra and Aqua have been providing data on global snow coverage. The daily data provided by the National Snow and Ice Data Center (NSIDC) (currently in version 6.1) is already of very high quality and serves as input data for the DLR Global SnowPack processor. There, remaining data gaps (e.g., due to clouds or polar night) are filled in four interpolation steps and cloud-free data is thus provided daily. So far this has only happened retrospectively after the end of a hydrological year (end of the meteorological summer). This data is now made available daily in near real-time with a time lag of 3 days in the EOC GeoService Portal. This enables the use of this data in time-critical issues, such as with regard to flood hazards. At the conference we will present the product, the access options, and possible applications.

[1] German Aerospace Center, Oberpfaffenhofen, Germany

An Analysis of Snowpack Temperature, Density, and Cold Content across the US West from the Repurposed USGS RMS Dataset

JEFFREY SCHMIDT[1], GRAHAM SEXSTONE[2], AND MARK SERREZE[1]

ABSTRACT

Snowpit records provide insight into snowpack conditions and changes, but many existing records are still in paper format, making analysis difficult. Undergraduate students digitized 1,818 historical, hand-written snowpit records from the Western U.S. Rocky Mountains. These unique vertical temperatures and density profiles enable analysis and physical understanding of snowpack cold content (CC) and other snowpack variables. CC represents the energy required to raise the snowpack's temperature to its melting point (related to the snowmelt onset date). It can be an indicator of large-scale climate change and is valuable for assessing the performance of numerical models of the internal snowpack energy and mass flux system. Digging snowpits and recording vertical temperature and density profiles from which temperature, density, and CC can be calculated is laborious and time-consuming. Few campaigns collect snowpit data and even fewer open-source, large geographic scale, high-quality datasets. For 29 years, the USGS Rocky Mountain Regional Snowpack Chemistry Monitoring Study (USGS RMS) has collected snowpit records at a network of 192 sites. At present, the record is updated annually at 60 snowpits. This thesis describes the compilation of the USGS RMS snowpit records into a quality-controlled dataset of temperature and density profiles followed by a detailed analysis of snowpack CC, temperature, density, and their variability across Rocky Mountain West.

This thesis demonstrates several important aspects of the dataset. First, the manual measurements are high quality and provide useful information to study the Rocky Mountain West snowpack and water resources. Snow water equivalent (SWE) contributes to the CC magnitude, but largely, temperature determines CC variability. CC is useful because it combines temperature and SWE into a single variable. CC reduces the internal snowpack physical system down to MJ m^{-2}, a basic energy term.

[1] Department of Geography, University of Colorado Boulder, Boulder, CO, USA
[2] USGS Colorado Water Science Center, Lakewood, CO, USA

Modeling Snowpack Bulk Density using Snow Depth, Cumulative Degree Days, and Climatological Predictor Variables

ANDRAS SZEITZ[1] AND ROBERT DAN MOORE[2]

ABSTRACT

Snowpack water equivalent (SWE) is a key variable for water resource management in snow-dominated catchments. While it is not feasible to quantify SWE at the catchment scale using either field surveys or remotely sensed data, technologies such as airborne LiDAR support the mapping of snow depth at scales relevant to operational water management. To convert snow depth to water equivalent, models have been developed to predict SWE or snowpack density based on snow depth and additional predictor variables. This study builds upon previous models that relate snowpack density to snow depth by including additional predictor variables to account for (1) the effect of annually varying meteorological conditions on densification through a cumulative degree-day index derived from North American Regional Reanalysis products, and (2) long-term climatologies that describe the prevailing conditions influencing regional snowpack properties. A non-linear model was fit to 114,506 snow survey measurements spanning 41 years from 1,166 snow courses across western North America. Under spatial cross-validation, the predicted densities had a root-mean-square error of 47.1 kg m^{-3}, a mean bias of -8 x 10^{-3} kg m^{-3}, and a Nash-Sutcliffe Efficiency of 0.70. Compared to a similar regression-based model reported in the literature, the two models had similar overall performance, but the model developed in this study had reduced seasonal biases. When applied to predict SWE from simulated depths determined from aerial platforms (LiDAR or Structure-from-Motion), 50% of the SWE estimates for May and June fell within -55 to 59 mm of the observed SWE, representing prediction errors of -15 to 23%.

[1] School of Earth, Environment and Society, McMaster University, Hamilton, ON, Canada
[2] Department of Geography, University of British Columbia, Vancouver, BC, Canada

Soil Freeze-Thaw Detection using Sentinel-1 SAR Data in Agricultural Fields

SHAHABEDDIN TAGHIPOUR JAVI[1], CHRISTOPHE KINNARD[1], AND ALEXANDRE ROY[1]

ABSTRACT

Seasonal soil freeze-thaw (FT) is a process of the flow of material and energy which occurs in the topsoil and proceeds into the deeper soil layers. In southern Canada, where the farmlands are dominantly located, the process plays an important role in hydrological regime, biogeochemical processes, and crop production in farmlands. To detect the FT state, the Saint-Maurice areas in southern Quebec has been selected as the test case for fieldwork purposes. This site has been instrumented and monitored to properly represent the heterogeneous characteristics of soil temperature in an agricultural field. We have used temperature loggers to measure soil temperature in eight patches using 40 points on two vertical levels (e.g., near-surface and 10 cm) including five patches in different agricultural fields and three in the forest areas. C-band synthetic aperture radar (SAR) of Sentinel-1 Ground Range Detected (GRD) scenes have been processed using Google Earth Engine (GEE) to generate the calibrated and incidence angle-corrected product. In this research, logistical regression of the freezing status against the freeze-thaw index has been applied. The FT index has been chosen during a post-harvest, frost, and frost-free period to get reference frozen and thawed values. Fitting a generalized linear model (GLM) as a flexible generalization of linear regression model shows that the probability of freezing in relation to the freeze-thaw index varies within and between farmland patches and forest patches. Based on the results, soil types and crop types significantly affect the probability of soil freezing in relation to Sentinel-1 backscattering in agricultural fields.

[1] Département des science de l'environnement, Université du Québec à Trois-Rivières, Trois-Rivières, QC, Canada

Fine-Scale Characterization of Snowpack Evolution using Unpiloted Aerial System Lidar and SfM Photogrammetry

MEGAN VERFAILLIE[1], EUNSANG CHO[2], JENNIFER JACOBS[1], ADAM HUNSAKER[3], FRANKLIN SULLIVAN[3], MICHAEL PALACE[4], ELIZABETH BURAKOWSKI[4], AND CAMERON WAGNER[3]

ABSTRACT

Unpiloted aerial systems (UAS) equipped with lidar and structure-from-motion (SfM) photogrammetry have emerged as viable methods to map snow depths at fine resolutions. In this study, eight UAS lidar/SfM snow depth maps were collected during the 2020/2021 winter season in Durham, NH, USA with the objectives of: (1) validating UAS snow depth retrievals using multiple *in situ* measurement techniques (Magnaprobe and field cameras), (2) conducting a quantitative comparison of lidar and SfM snow depths (< 35 cm) throughout the winter, and (3) better understanding a spatial structure of snow depth and its relationship with terrain features. The surveys were conducted over approximately 0.35 km^2 including large open and mixed forested areas. Results showed that lidar outperformed SfM compared to *in situ* observations, especially at forested sites. In the field, lidar had a lower mean absolute error (MAE) than SfM compared to the Magnaprobe (lidar = 3 cm, SfM = 5 cm) and field cameras (lidar = 3 cm, SfM = 14 cm). The same was true in the forest compared to the Magnaprobe (lidar = 7.2 cm, SfM = 32 cm) and cameras (lidar = 2.7 cm, SfM = 45 cm). The difference between lidar and SfM was much greater in the forest, with SfM overestimating snow depths at most locations. We also found differences between the *in situ* measurement techniques used for validation. However, the differences had only a modest impact on snow depth validation. The spatial structures of snow depth captured by lidar were generally consistent throughout the period indicating that static land characteristics such as slope, vegetation, and soil properties, may control the spatial variability.

[1] Department of Civil and Environmental Engineering, University of New Hampshire, Durham, NH, USA
[2] NASA Goddard Space Flight Center, Greenbelt, MD, USA
[3] Institute for the Study of Earth, Oceans, and Space, University of New Hampshire, Durham, NH, USA
[4] Department of Earth Sciences, University of New Hampshire, Durham, NH, USA

Precipitation Phase from Atmospheric Model Improves Snowfall Estimates across Canada

VINCENT VIONNET[1], MARC VERVILLE[1], VINCENT FORTIN[1], FRANCOIS LEMAY[1], MELINDA M. BRUGMAN[1], AND JULIE THÉRIAULT[2]

ABSTRACT

The precipitation phase at the surface strongly affects the evolution of the snowpack and the associated hydrological response in cold regions. Snowpack models deployed at continental scales usually rely on ground-based precipitation-phase partitioning methods (PPMs) that use near-surface air temperature and humidity. These PPMs still ignore atmospheric conditions in the layers aloft, thus limiting their ability to predict the precipitation phase at the surface.

In this study, gridded phase estimates from a variety of ground-based and atmospheric-based PPMs are evaluated against manual observations of precipitation phase collected across a large region covering Canada and the northern United States (US) from September 2019 to June 2020. The hourly gridded phase estimates are derived from the High Resolution Deterministic Prediction System (HRDPS) running at 2.5-km grid spacing over the region of interest. Two atmospheric-based PPMs are considered from HRDPS: (i) the Bourgouin method relying on the vertical profile of air temperature and (ii) the Latent Heat Release Method (LHRM) relying on the vertical profile of wet-bulb temperature. Ground-based PPMs of various complexity are also considered: (i) single air temperature thresholds, (ii) snowfall fraction depending on air temperature, and (iii) humidity-based PPMs.

Results show that humidity-based PPMs provides the best estimate of precipitation phase at continental scales among the ground-based PPMs. In particular, they outperform PPMs relying only on air-temperature in the mountainous regions of western Canada and US. In these regions, LHRM offers the best performances thanks to its ability to forecast the height of the snow level. In eastern Canada and US, humidity-based PPMs tends to overestimate the occurrence of snowfall. Atmospheric-based PPMs improves these estimates since they can account for the presence of a melting layer above the surface, often encountered during precipitation events in these regions. These results encourage the use of atmospheric-based PPM for snowpack modeling across Canada.

---

[1] Environment and Climate Change Canada, Dorval, QC, Canada
[2] Département des sciences de la Terre et de l'atmosphere, Université du Québec à Montréal, Montréal, QC, Canada

Estimating Snow Water Equivalent at the Watershed Scale using Drones in the Arctic Shrub-Tundra

BRANDEN WALKER[1], ROSAMOND TUTTON[1], AND PHILIP MARSH[1]

ABSTRACT

Spatial variations in snow depth and density translate to heterogenous snow water equivalent (SWE) in shrub-tundra environments. Spatial variability in snow depth is primarily due to winter blowing while density variability is due to snow metamorphism process. Recent advances in remotely piloted aircraft systems (RPAS) and structure-from-motion photogrammetry (SfM) have allowed for the measurement of high-resolution spatial and temporal changes in snow depth across snow covered landscapes. This is a major advance in high-resolution remote sensing that provides significant opportunities to advance hydrological process understanding and testing of predictive snow models. However, due to complexities relating to accurately estimating similar spatial or temporal variability in snowpack density, there has been little advance towards mapping SWE at comparable scales. In this paper we address these shortcomings by combining high-resolution snow depth maps created using RPAS SfM with spatially distributed snowpack bulk density observations to accurately map SWE for a small (1 km^2) watershed. These data allowed mapping of late winter snow depth, density and SWE across dominant vegetation and topographic land cover types. Spatially distributed snowpack density observations were collected using representative snow surveying and snow pits. An empirical relationship between snow depth and density was produced and applied to the SfM snow depth maps to estimate SWE. Final SWE maps were validated using *in situ* snow depth and SWE observations obtained within the study area, demonstrating highly accurate snow depth and SWE results. Furthermore, we were able to provide an in-depth analysis of late winter SWE variability across the dominant landcover types, revealing significant differences in water storage across the watershed.

[1] Department of Geography and Environmental Studies, Wilfred Laurier University, Waterloo, ON, Canada

An Observing Simulation System Experiment (OSSE) for Snow Mass Estimation over Western Colorado using Adaptive Viewing from Space

LIZHAO WANG[1], COLIN MCLAUGHLIN[1], BARTON FORMAN[1], SUJAY KUMAR[2], PAUL GROGAN[3], RHAE SUNG KIM[2], AND MELISSA WRZESIEN[2]

ABSTRACT

Snow plays a key role in the freshwater supply of many basins with headwaters in mountainous regions. Optical LiDAR is a promising technology that can help improve our knowledge of snow mass. However, optical LiDAR viewing, in general, is significantly constrained by swath width limitations and cloud attenuation. One potential technique to ameliorate this issue is to dynamically slew the space-based sensor so that it preferentially views the target of interest, and as a result, observes the target of interest (i.e., snow-covered land) more often relative to a traditional, fixed viewing approach.

In this study, we develop an observing simulation system experiment (OSSE) in order to explore the trade-off space between a strategy of fixed sensor viewing versus adaptive sensor viewing. In the experiment, we first generate a "synthetic truth" of snow depth using the NoahMP-4.0.1 land surface model within the NASA Land Information System (LIS). Afterwards, synthetic snow depth retrievals for a given LiDAR configuration are generated using both fixed and adaptive viewing strategies, respectively. Next, a realistic amount of observation error is then injected into the synthetic retrievals so that they serve as a reasonable proxy for real-world LiDAR retrievals. The synthetic retrievals are then assimilated into NoahMP-4.0.1 using an ensemble Kalman filter (EnKF) in order to help quantify the added value that each type of snow retrieval has on the land surface model performance. The results from this OSSE help reveal the advantages (and disadvantages) of adaptive viewing strategies, and in turn, can help mission planners in determining how to get the most observational "bang for the buck" in the selection of a future snow mission.

[1] Department of Civil and Environmental Engineering, University of Maryland, College Park, MD, USA
[2] NASA Goddard Space Flight Center, Greenbelt, MD, USA
[3] School of Systems and Enterprises, Stevens Institute of Technology, Hoboken, NJ, USA

Characterizing the Role of Snow for Liquid Water Storage and Transmission: A Ground-Based Remote Sensing and Modeling Sensitivity Analysis

RYAN WEBB[1,2], KEITH MUSSELMAN[2], SIOBHAN CIAFONE[3], KATHERINE HALE[2,3], AND NOAH MOLOTCH[2,3]

ABSTRACT

Streamflow response in headwater catchments is highly sensitive to the hydrologic connectivity of hillslopes to streams during spring snowmelt. Despite strong evidence at point- to plot-scales of flow paths creating lateral connectivity within an alpine snowpack, meltwater is commonly assumed to infiltrate vertically through the snowpack. Hydrologic models only treat the horizontal (downstream) routing of water once released from the snowpack and/or soil column. This assumption limits our ability to represent the full dynamic nature of hydrologic connections in snow-dominated mountainous headwaters. Thus, the goal of this study is to assess the mechanisms that control the spatiotemporal distribution of liquid water in an alpine snowpack during the spring snowmelt season. We utilize terrestrial laser scanning (TLS), ground penetrating radar (GPR), and manual observations to map the seasonal dynamics of snow depth, snow water equivalent (SWE), and within-snow liquid water content (LWC). We compare these observations to point-scale parameter sensitivity analyses with a modular snow model (SUMMA). The results show high spatial variability of LWC in an alpine snowpack during snowmelt. Statistical analyses show LWC is most highly correlated to snow depth (coefficient of determination = 0.62). However, including the distance to bare soil and topographical slope slightly improved the coefficient of determination (coefficient of determination = 0.67). While hydrologic models have the flexibility to simulate many of the observed dynamics in snowpack liquid water storage, model simulations using previously published parameter ranges always underestimated the high liquid water storage at one of the three sites. This is likely a result of current model structures that lack capabilities for surface ponding of water within a snowpack or surface runoff laterally through a snowpack. Our slope-scale characterization of the spatiotemporal distribution of in-snow LWC, together with a model-based sensitivity assessment, will inform future efforts in hydrologic model development and catchment observations.

[1] Department of Civil and Architectural Engineering, University of Wyoming, Laramie, WY, USA
[2] Institute of Arctic and Alpine Research, University of Colorado Boulder, Boulder, CO, USA
[3] Department of Geography, University of Colorado Boulder, Boulder, CO, USA

Merging Models with Observations to support Open Science, NASA SnowEx, and Snow Satellite Missions

MELISSA WRZESIEN[1], SUJAY KUMAR[1], CARRIE VUYOVICH[1], RHAE SUNG KIM[1], AND EUNSANG CHO[1]

ABSTRACT

Seasonal snow accumulation and melt have a critical role in global water and energy budgets, and the 2017 Earth Science Decadal Survey recommended snow water equivalent (SWE) as an "explorer priority" for future missions. In recent years, NASA SnowEx field campaigns have tested multiple remote sensing techniques as potential mission concepts by collecting ground and airborne observations of snow across a range of land cover and snow types. However, a single sensor will not be able to provide estimates of all snow types in all conditions globally; instead, an integrative approach is required, utilizing multiple observational sources and merging them with models for complete spatiotemporal coverage.

To address the need for an observation-model merging environment for improving estimates of snow properties such as SWE, snow density, snow grain size, and albedo, we discuss plans for the upcoming 2023 SnowEx Alaska campaign. Field and modeling efforts will be brought together, including running near real-time model simulations with the NASA Land Information System (LIS) and designing assimilation experiments once field observations have been collected. We also present ongoing efforts with the NASA Earth Information System (EIS), a project focused on open science and improving the accessibility of data. Discussions will include a pilot study of extreme snowmelt flooding in the Midwest U.S. and how EIS efforts can help build an observation-model environment for snow that addresses the need for open-source science.

Such data integration approaches not only benefit field campaigns but merging model and field efforts will also be a crucial step in a future snow mission. This presentation will include a discussion on how the SnowEx model-data fusion work can inform future plans for a snow satellite mission and how such efforts should meet NASA's commitment to open science.

[1] NASA Goddard Space Flight Center, Greenbelt, MD, USA

Evaluation of Three Different Machine Learning Algorithms for Snow Mass Estimation over the Colorado Rockies using Space-Based Passive Microwave Brightness Temperatures

BINCHENG YU[1] AND BARTON FORMAN[1]

ABSTRACT

This study compares the performance of three different machine learning algorithms used for snow water equivalent (SWE) estimation. Inputs to these algorithms include passive microwave (PMW) brightness temperature (Tb) observations at 10.65 GHz, 18.7 GHz, and 36.5 GHz at both vertical and horizontal polarization as collected by the Advanced Microwave Scanning Radiometer (AMSR-2). The three algorithms include: 1) support vector machine (SVM) regression; 2) long short-term memory (LSTM) networks; and 3) Gaussian process (GP) regression. *In situ* SWE measurements from the SNOTEL network collected across western Colorado are used as the training "targets" during the training procedure. The performance of the algorithms is evaluated using a number of different metrics including, but not limited to, root mean square error (RMSE), correlation coefficient, mean square error (MSE), and bias. The evaluation is conducted over a range of different elevations and different land cover classifications in order to assess algorithm performance across a broad range of snowpack conditions. Preliminary results suggest the LSTM algorithm is computationally more efficient during the training process as compared to the other algorithms yet yields a similar level of performance. However, some limitations have been found in the study, including poor performance during deep snow conditions, which is likely related to signal "saturation" within the PMW Tb's used during the supervised training process. Additionally, algorithm performance is strongly dependent on the amount of training data such that too little training data results in poor performance by the algorithm at successfully reproducing inter-annual variability. The strengths and limitations of these different machine learning algorithms for snow mass estimation will be discussed.

[1] Department of Civil and Environmental Engineering, University of Maryland, College Park, MD, USA

Session #5: Observation and Modeling of Snow Processes: Integrating Legacy and New Tools to Advance Snow Science

Assessing the Spatial Distribution of Snow Depth in Open and Forested Environments by UAV Lidar

VASANA DHARMADASA[1], CHRISTOPHE KINNARD[1], AND MICHAEL BARAËR[2]

ABSTRACT

Spatial distribution of the snow depth in two agro-forested sites and one coniferous site in eastern Canada was analyzed for topographic and vegetation effects on snow accumulation. Spatially distributed snow depth retrievals were obtained by Unmanned Aerial Vehicle Light Detection and Ranging (UAV lidar) measurements in 2019 and 2020. Lidar-derived snow depths showed distinct patterns of snow accumulation and erosion in open areas/fields versus adjacent forested areas across all sites. Omnidirectional semi-variogram analysis of snow depths showed the existence of a scale break distance less than 10 m in the forested area at all three sites whereas open areas showed scale invariance or comparatively larger scale break distances (i.e., 18 m). To investigate the effect of vegetation and topographic variables on the spatial variability of the snow depth, random forest models were employed. Our results showed that including wind-related forest edge effects improved the model accuracy by more than 50% in agro-forested sites whereas incorporating canopy characteristics improved the model accuracy by more than 60% in the coniferous site. This implies the importance of including and better representing these processes in process-based models for accurate estimates of snowpack dynamics. As well, within agro-forested sites, it is rather the underlying topography and/or the wind-redistribution of snow along forest edges that govern the snow depth variability, while within the coniferous environment, it is the forest structure variability. The findings of this study could be applied/extrapolated to similar landscapes in the region and any similar environment that addresses the relevant processes.

[1] Département des science de l'environnement, Université du Québec à Trois-Rivières, Trois-Rivières, QC, Canada
[2] Department of Construction Engineering, École de technologie supérieure, Montreal, QC, Canada

Simple Snow Temperature Index Models Account for Important Discrepancies between Snow Water Equivalent Products

ALEKSANDRA ELIAS CHEREQUE[1], PAUL KUSHNER[1], CHRIS DERKSEN[2], AND LAWRENCE MUDRYK[2]

ABSTRACT

Snow water equivalent (SWE) is a routine output from reanalysis systems and offline land-surface schemes, but documented discrepancies among the resulting SWE products impede the assessment of hemispheric-scale snow mass variability and trends. These discrepancies may stem from differing meteorological fields between reanalysis products or different surface observation sources, data assimilation schemes, snow model structure, and parameterizations used.

In this work, we investigate SWE differences between three modern reanalysis products (ERA5, JRA-55, and MERRA-2) from 1980-2020. We force the recently updated B-TIM, a simplified, single-layer temperature index snow model based on Brown *et al.*, (2003), with their respective temperature and precipitation fields. B-TIM captures the most important features of snow evolution, whereby precipitation amount and phase mainly influence the resulting snow depth, while temperature influences the snowpack density, but it does not attempt to represent humidity-, radiative-, or wind-related processes. Nonetheless, by standardizing the snow model physics used, we primarily isolate biases in the precipitation and temperature fields as sources of spread.

We identify and correct first-order biases between the JRA-55 and ERA5 forcing data to bring the two B-TIM reconstructed datasets, ERA5rec and JRArec, into closer agreement. This reduces the climatological mean wintertime bias by 50%, indicating the propagation of forcing biases into SWE uncertainty. Through several SWE metrics at the continental scale (e.g. mean SWE, max SWE), we find strong correlations in the range 0.84 to 0.96 between all the reconstructed datasets. This is not observed between the native SWE products, which have correlations below 0.4, suggesting that snow model and data assimilation differences affect the interannual variability of a given product more than relative forcing biases do. We conclude that comparison to a temperature index model helps diagnose some inconsistencies between snow products and help assess the impact of changes in data assimilation schemes and input data.

[1] Department of Physics, University of Toronto, Toronto, ON, Canada
[2] Environment and Climate Change Canada, Toronto, ON, Canada

Improving Snow Analyses for Hydrological Forecasting at Environment and Climate Change Canada (ECCC)

CAMILLE GARNAUD[1], VINCENT VIONNET[1], ÉTIENNE GABORIT[1], VINCENT FORTIN[1], MARCO CARRERA[1], BERNARD BILODEAU[1], AND DOROTHY DURNFORD[1]

ABSTRACT

As part of the National Hydrological Services Transformation Initiative, Environment and Climate Change Canada (ECCC) designed and implemented the National Surface and River Prediction System (NSRPS) to provide surface and river flow analysis and forecast products across Canada. Within NSRPS, the Canadian Land Data Assimilation System (CaLDAS) produces snow analyses that are used to initialize the land surface model, which in turn is used to force the river routing component. Originally, CaLDAS was designed to improve atmospheric forecasts with less focus on hydrological processes. When snow data assimilation occurs, the related increments remove/add water from/to the system, which can sometimes be problematic for streamflow forecasting, particularly during the snowmelt period. Recent modifications to the snow analysis method introduce multiple innovations that respond to the need for higher quality snow analyses for hydrological purposes, including the use of IMS snow cover extent data instead of *in situ* snow depth observations. Results show that the new snow assimilation methodology brings an overall improvement to snow analyses and substantially enhances water conservation, which is reflected in the generally improved streamflow simulations. This work represents a first step towards a new snow data assimilation process in CaLDAS, with the final objective of producing a reliable snow analysis to initialize and improve NWP as well as environmental predictions, including flood and drought forecasts.

[1] Environment and Climate Change Canada, Dorval, QC, Canada

Complex Precipitation Phase and Impact on Snowpack Evolution in Eastern Canada

NICOLAS LEROUX[1], VINCENT VIONNET[2], JULIE THÉRIAULT[1], HADLEIGH THOMPSON[1], DOMINIQUE DROLET[1], LISA RICKARD[3], STEPHEN DÉRY[3], AND RONALD STEWART[4]

ABSTRACT

Accurate estimation of precipitation phase at the surface is critical for hydrological modelling in cold regions. Different precipitation phase partitioning methods (PPM) exist and were developed with local data; however, their application across a variety of climatic regions in numerical snow models is often limited. In particular, PPMs used in hydrological and land surface models struggle to represent near-0 °C complex precipitation phase, such as freezing rain and wet snow, and their impact on snowpack simulations. The goal of this study is to evaluate PPMs of varying complexity using high-quality observations of precipitation phase and to assess the impact on simulated snowpack evolution over a winter season. To do so, we use meteorological data including air temperature, relative humidity, wind speed, and precipitation amount collected at Edmundston, New Brunswick, during the Saint John River Experiment on Cold Season Storms field campaign. These data are combined with observations of snow on the ground, such as snow depth, snow surface temperature, and snow water equivalent. The reference precipitation phase is derived from measurements from a laser-optical disdrometer, micro-rain-radars (MRR), and a Multi-Angle Snowflake Camera (MASC) operating during SAJESS. The snowpack evolution during the 2020-2021 winter was simulated using the model Crocus and the impact of the different PPMs on simulated snowcover dynamics was evaluated across the winter season. The results show that commonly used PPM are limited when representing wet snow and freezing rain events, which mostly impacted the snowpack evolution in spring due to a succession of storms associated with complex precipitation phase. Overall, this study highlights the difficulty in estimating precipitation phase at the surface during near 0 °C conditions and its impact on snow modeling.

[1] Département des sciences de la Terre et de l'atmosphère, Université du Québec à Montréal, Montréal, QC, Canada

[2] Environment and Climate Change Canada, Dorval, QC, Canada

[3] Department of Environmental Science, University of Northern British Columbia, Prince George, BC, Canada

[4] Clayton H. Riddell Faculty of Environment, Earth, and Resources, University of Manitoba, Winnipeg, MB, Canada

Operational Water Forecast Assessment of a Spatially Distributed Process-Based Snow Model: A Case Study in the East River Watershed, Colorado

JOACHIM MEYER[1], MCKENZIE SKILES[1], PATRICK KORMOS[2], ANDREW HEDRICK[3], ERNESTO TRUJILLO-GOMEZ[3], AND SCOTT HAVENS[3]

ABSTRACT

Operational water-resource planning faces an increased challenge with a changing seasonal snowpack in mountain watersheds due to global and regional climatological factors. An example region is the western United States, where there is a demonstrated decline in extent and amount of seasonal snow in mountain ranges such as the Sierra Nevada, California, or the Rocky Mountains, Colorado. Causes for the shift include precipitation phase changes or increased amounts of dust on snow. Like the Colorado Basin River Forecast Center (CBRFC), regional forecasters cannot currently account for these factors when their prediction method relies on an empirical snow model based on historic calibration records. To evaluate the options and supplement the current method of the CBRFC, we run a physical-based snow energy balance model for past water years in a subset region: the East River Watershed, Colorado. The results are compared with *in situ* measurements, remote sensing observations, and the predictions by the current model. This assessment is an effort to include the process-based model in day-to-day CBRFC operations and to create a foundation to expand to larger domains. This project also bridges the gap between scientific advancements and benefits for society with more accurate water resource forecasting.

[1] Department of Geography, University of Utah, Salt Lake City, UT, USA
[2] Colorado Basin River Forecast Center, NOAA National Weather Service, Salt Lake City, UT, USA
[3] Northwest Watershed Research Center, USDA Agricultural Research Service, Boise, ID, USA

Simulation of Cold Processes in the CMIP6 Land-Historical Simulations

LAWRENCE MUDRYK[1], ELEANOR BURKE[2], GERHARD KRINNER[3], NATE COLLIER[4], CHRIS DERKSEN[1], AND DAVID LAWRENCE[5]

ABSTRACT

Model evaluation is a necessary component of climate research. While the snow and soil components of land-surface models have been previously evaluated using standalone simulations (Dirmeyer *et al.*, 1999; Slater *et al.*, 2001; Dirmeyer, 2011), the Land Surface, Snow and Soil moisture Model Intercomparison Project (LS3MIP; van den Hurk *et al.*, 2016) is the first CMIP framework that coordinates both components and integrates them into the larger suite of CMIP experiments. Here we analyze output from the 'land-hist' global land surface simulations, forced by historically observed meteorological conditions, and compare them to the fully coupled and AMIP historical CMIP6 simulations.

We focus on high-latitude processes and variables, in particular snow and permafrost. For both snow and permafrost there is better representation of trends and historical variability in the 'land-hist' simulations. NH snow mass and snow extent variability and trends correlate better with observations in the historically forced simulations. Likewise, these simulations better represent historical permafrost loss rates for nearly all analyzed models.

However, there remain residual structural and parametrization errors in both variables. While for alpine snow mass, the 'land-hist' ensemble mean shows better agreement with observations and smaller errors for individual models (reduced spread about the observations) this is not the case for non-alpine snow, which has larger systematic errors relative to the observations. We analyze these differences in terms of the balance of accumulation and ablation across both regions. Similarly for permafrost, errors in the functional relationships that control permafrost extent and soil temperature are large and result in limited to no improvement in climatological permafrost extent in the historically forced simulations. These results suggest that the added value of standalone land model simulations in the suite of CMIP6 experiments may depend on both the variable and region of interest.

[1] Environment and Climate Change Canada, Toronto, ON, Canada
[2] Met Office Hadley Centre, Exeter, UK
[3] Institut des Géosciences de l'Environnement, Université Grenoble Alpes, Grenoble, France
[4] Oak Ridge National Laboratory, Oak Ridge, TN, USA
[5] National Center for Atmospheric Research, Boulder, CO, USA

SNO-FOO 2022 – Not awarded

Past Recipients

The Eastern Snow Conference annually bestows upon a distinguished snow scientist who, in striving for excellence in snow research, contributes to an event of notable humor the highly coveted Sno-Foo Award.

Year	Winner	Affiliation
2022	*Not awarded*	
2021	*Not awarded*	
2020	*Meeting cancelled*	
2019	Alexandre Roy	Université du Québec à Trois-Rivières
2018	Alexandre Langlois	Université de Sherbrooke
2017	Michael Durand	The Ohio State University
2016	Roger De Roo	University of Michigan
2015	Kevin Côté	Université de Sherbrooke
2014	Dorothy Hall	NASA GSFC
2013	Benoit Montpetit	Université de Sherbrooke
2012	Don Pierson	NYC DEP
2011	Ken Rancourt	Mt. Washington Observatory
2010	Kyung-Kuk (Kevin) Kang	University of Waterloo
2009	Rob Hellström	Bridgewater State University
2008	Steven Fassnacht	Colorado State University
2007	The Group of 9[1]	
2006	Andrew Klein	Texas A&M University
2005	Claude Duguay	University of Alaska-Fairbanks
2004	Chris Derksen	Meteorological Service of Canada
2003	Miles Ecclestone	Trent University
2002	Danny Marks	U.S.D.A.
2001	Brenda Toth	University of Saskatchewan
2000	Mauri Pelto	Nichols College
1999	Ross Brown	Meteorological Service of Canada
1998	Mary Albert	CRREL
1997	Jean Stein	Université de Laval
1996	Colin Taylor	Trent University
1995	Mike Demuth	NHRI
1994	Bert Davis	CRREL
1993	John Pomeroy	NHRI
1992	Tom Niziol	National Weather Service
1991	Terry Prowse	NHRI
1990	Kersi Davar	University of New Brunswick
1989	Gerry Jones	INRS-EAU
1988	Robert Sykes	SUNY
1987	John Metcalfe	Meteorological Service of Canada
1986	Peter Adams	Trent University
1985	Don Wiesnet	National Weather Service
1984	Barry Goodison	Meteorological Service of Canada

[1] Jimmy MacDonald (University of Saskatchewan), Bill Floyd (University of British Columbia), Chris DeBeer (University of Saskatchewan), Wendell Koning (Government of Alberta), Jaime Hood (University of Calgary), Danika Muir (University of Calgary), John Jackson (University of Calgary), Sarah Forte (University of Calgary), Masaki Hayashi (University of Calgary)